Gaiome
Notes on Ecology, Space Travel and Becoming Cosmic Species

Kevin Scott Polk

Gaiome: Notes on Ecology, Space Travel and Becoming Cosmic Species
Copyright © 2007 Kevin Scott Polk.

ISBN 978-1-60145-242-9

Printed in the United States of America.

Computations, typesetting and interior illustrations by the author using Ruby, LATEX and Star Office.

Gaiome.com
Bloomington, Indiana
2007

Contents

List of Figures

Acknowledgments

What will it *really* take to get into space? For over 20 years, I've been posing this question to astronomers, rocketeers, science program managers and others I've met in the space business. In every case, their eyes betrayed an instant recognition of what I was asking. This field is haunted by a sense of unfulfilled promise: a canyon in the heart that has widened each day since astronauts last walked the Moon. Still, most of these specialists quickly recovered their composure and waved the question away. I am deeply grateful, therefore, to those few who would talk candidly and thoughtfully about their hopes, fears and dreams for a grand human future in space. For their gracious discussions and correspondences, I thank Professor Freeman Dyson, Dr. Jeffrey F. Bell, Astronaut Stanley G. Love, Ph.D., Dr. Jancy McPhee, Astronaut Edwin "Buzz" Aldrin, Ph.D., Dr. Alan Binder, Jeff Greason, Jim Benson, and Professor John S. Lewis. For years of instruction and insight into NASA's labyrinthine political lore, I thank Dr. B. Ray Hawke. For their long friendships and many space-related discussions, I thank Dr. Charles J. Budney, Rex Ridenoure and Dr. Moriba Jah. This book does not reflect their views (except where specifically noted), only their encouragement to work the problem.

I am grateful to my parents, Dr. Barbara Polk and Dr. Lon Polk, for a lifetime of encouragement and for comments on two wildly different drafts of this book. Thanks also go to fellow permaculturist and author Dr. Ann Kreilkamp, who pointed out at a critical juncture that I'm not so much telling a story as inventing a language: a new way to talk about space and our place in it.

For many years of discussion and reading assignments in the world of business, I thank Von Kenrick Kaneshiro. Thanks also go to my mentors in ecology and sustainability: Peter Bane and Keith Johnson, whose 80-hour Permaculture Design course opened many new doors; Rhonda Baird, who established the Bloomington Permaculture Guild; Michael Simmons and Lucille Bertuccio, who recruited an impressive array of experts to co-teach Bloomington's 13-week Grow Organic Educator Series and 30-hour Wildlife Habitat Stewards training course; and Brook Gentile, whose weekly volunteer gardening group with Mother Hubbard's Cupboard is gradually teaching me how to grow.

Several of my researches began with wikipedia.org, an ecology of learning that democratizes not only knowledge, but the process of improving it. In gratitude, I now donate my monthly reading budget twice a year to the Wikimedia foundation. Props also go to the many artists at Magnatune.com, who serenaded me throughout this project.

Kudos to Angela Hoy at Booklocker.com, who brought this book into print with dazzling speed, professionalism and cheer.

Most of all, I thank my wife, Dr. Kimberly Wagner, whose love, support and belief have made everything possible.

Introduction

On a crisp fall evening in 1985, the Princeton University chapter of Students for the Exploration and Development of Space (SEDS) held an open meeting. About two dozen undergraduates attended, spreading out in groups of two or three among the raised, hardwood pews of a small lecture hall.

The club's President, Jeff Bezos, talked a little about SEDS and a lot about his dreams for a glorious future in space. At one point he was describing a scheme to build gigantic space habitats that theoretically could house millions of people. The construction technique involved using huge solar mirrors to heat a metal asteroid until it was completely molten. Then workers would plunge a long tungsten tube into its center and inject large quantities of water. This would flash into steam, inflating the asteroid like a balloon to make a spherical hull—

A loud slam cut him off. A student in the middle of the room jumped up and, choking back sobs of rage, yelled, "How dare you rape the universe!" After she had stormed out, Jeff, more bemused than ruffled, leaned toward me and another SEDS officer and said: "Did I hear her right? Did she really just defend the inalienable rights of *barren rocks?*"

Jeff, like the rest of us, had no trouble brushing aside criticism before it could sink in. Having grown up in the belt-tightening Carter administration, under the Cold War threat of Mutually Assured Destruction, Earth seemed small to us: fragile and crowded. Beyond lay boundless space, with limitless energy, resources and opportunity. Out There, population and consumption, and hence science and the arts, could grow forever. Our template was Gerard K. O'Neill's *The High Frontier* (1976), a plan to build miles-long habitats along the inside walls of huge, spinning cylinders in orbit.[1] Green and spacious, O'Neill colonies would house tens of thousands or even millions of people. The colonists would earn their way by building giant orbiting solar power stations

7

that would beam energy down to microwave receivers on Earth. Everything would be built using lunar and asteroidal ores, eventually moving all polluting industry off-Earth. According to O'Neill, the colonies could earn and grow fast enough to off-load Earth's entire population in a mere 35 years.[2] The Moon and asteroids had enough mineral resources to build thousands of Earths-worth of new, enclosed land. In the era of energy crises, *Limits to Growth* and Skylab, O'Neill's proposal made the front pages of *Physics Today*, *Science* and the *New York Times*, attracting a grass-roots following of techno- and eco-humanists that has not been rivaled since.

Still, a long and notable list of critics were as furious at the concept as Jeff's accuser. Historian Lewis Mumford regarded space colonies as "technological disguises for infantile fantasies." Nobel prize-winning biologist George Wald wrote "Let me say at once that I view them with horror." Educational reformer John Holt remarked that O'Neill and his followers had "lost the feel of real things."[3] To the critics, the whole mythology of finding escape in the heavens from the wreckage we seemed certain to make of Earth seemed both apocalyptic and futile. Given our struggles to sustain ourselves on Earth, where evolution and long experience have adapted us to its abundance, how could a single generation hope to do better in an unexplored, radioactive vacuum?

Privately, I had a few doubts of my own. Not about space colonies; the problem was the Space Shuttle. Still in the design stages when O'Neill first proposed his colonies, it was finally flying—at fifty times the ticket price that the National Aeronautics and Space Administration (NASA) had promised. Of course that killed O'Neill's colonies, which depended on low transportation costs. But NASA was the only game in town. Where else could spacers such as myself go to realize our dreams?

That fall, I sought out celebrity physicist-author Freeman Dyson for some career counseling. On the theory that the Henry Fords of rocketry would emerge from garages rather than NASA centers, he gently suggested that I get out of the space business entirely and earn my living in the much more lucrative field of computers. With some luck, perhaps I could earn enough money to experiment with new launch schemes as a hobbyist. I believed him, but I couldn't turn away—not even a few months later, when the Space Shuttle Challenger crashed. Instead, I went into astronomy, managed an archive of images from NASA's planetary missions, designed satellite orbits and parts, published astronomy software and trained to operate the Microphone experiment aboard the doomed Mars Polar Lander.

Meanwhile, cleverer souls followed Dyson's plan. In college, Jeff Bezos switched from aerospace to computer science. Later he founded Amazon.com, made billions, and launched a very secretive space company that, according to its web site (www.blueorigin.com), is "creating an enduring human presence in space." Bezos is hardly alone. In recent years, at least six self-made billionaires have begun to experiment with passenger space transport. Unlike Christopher Columbus or Nazi rocketeer Wernher von Braun, they don't need to convince skeptical rulers or dip into the national treasury; they already have the money. They can explore space any way they want. They know technology. They know business. They know what to do.

Or do they? Giving the old space dreams a new, corporate face would hardly comfort the critics. Will Bezos, Musk, Branson and the rest have any better "feel for real things" than NASA? Will they beat NASA's prices by factors of tens to hundreds? Will they make space launch safe enough to attract millions of travelers? And if they succeed, can humanity expand into space without also expanding its fatal wars on itself and nature, as the critics had warned?

With these questions in mind, I began building spreadsheet models of a rocket business. To inject some economic reality into the analysis, I started with a hefty operating profit margin and worked backward through the technical details to obtain a better estimate of ticket price (which rocketeers all too often and quite erroneously equate to their operating costs). The results were mixed. I found that a private company probably could send a passenger safely and profitably into orbit for $140,000 rather than today's going rate of $20 million. But the development costs, lifted straight from published figures in aviation, came to several times as much as the space entrepreneurs appear to be spending. Worse, many of them appear to be building the wrong kind of rocket and trying to sell it to the wrong customer.

Next I turned to O'Neill's cylindrical colony design—and recoiled as technical flaws leapt out of nearly every system. With an inherently unstable rotation, mirrors too large to hold their shape and a chemically volatile atmosphere, the design was a giant Rube Goldberg contraption just waiting to burn up or fly apart. O'Neill and colleagues had known about some of these issues, but angrily waved them away as engineering details. With a cadre of True Believers at NASA and elsewhere, the design and its economic basis in beamed solar power have remained largely unchallenged. Until now.

I began by simplifying the habitat design, making its hollow shape short and squat for stability and turning it on its side to avoid pointing problems. I got rid of the co-rotating mirrors, external shielding, agriculture pods, dish antennae and other protrusions. This made it easier and cheaper to build and maintain. As with an O'Neill colony, it would spin to simulate gravity, allowing people to live on its inside walls. I chose a thick hull so it could hold an Earth-like atmosphere at sea-level pressure. This also helped it resist radiation. To simplify the problem of recycling, I surveyed the biospherics literature for clues about relying less on untried mechanical systems and more on familiar plants and soils. This drew me deeper into ecology—and led to a paradox.

In the mid-1970s, Australian ecologists Bill Mollison and David Holmgren had developed *permaculture*, a practice of designing homes, towns and cities that sustained themselves through complex, forest-like ecologies. This may seem a step backward until you consider the enormous efficiency of forest systems. For example, acorns from an oak woodland can match a wheat field in terms of calories produced per acre. Yet unlike our monocrop agriculture, a natural forest includes many other plant species, all of which are edible—either by humans or hundreds of other animals.[4] Healthy forests can also have hundreds to thousands of times less soil erosion and dozens of times better nutrient recycling than monocrop agriculture.

As a designer, I could not ignore these efficiencies. How might permaculture work in space? I imagined, as Russian space pioneer Konstantin Tsiolkovsky had over a century ago, a "greenhouse conservatory" that could run on sunlight as autonomously as Earth itself.[5]

How odd, then, that O'Neill and his libertarian followers, who knew of Tsiolkovsky, would design their tiny world as a *colony*: an economic possession of a distant nation or corporation. Like colonial powers throughout history, these owners would have every incentive to secure and control their formidable assets by any means available, including debt bondage and coercive monopolies. About the last thing they would ever want to do is make the colonies autonomous. Thus the colonists themselves would have even less freedom than today's ground-controlled astronauts.

My notion of an autonomous, permacultural mini-world did not fit the colonial model. Lacking significant exports or need for imports, it would offer prospective investors little by way of recurring income. For residents, though, it would provide plenty of value as permanent real estate—the very thing that launched O'Neill and his students into space studies in the first place.

But if not a colony, what should I call it? Certainly not a *habitat*, which connotes problems long-since solved. The word *biosphere* fit, but it also fit everything from sealed glass bubbles with algae and brine shrimp to the entire Earth itself. Paolo Soleri's *arcology*, Dandridge Cole's *Macro-Life*, Isaac Asimov's *spome* and the Artemis Society's *xity* each described space dwellings, sometimes employing biological metaphors. But all of these schemes were urban and human-centered, housing only selected species as necessary to provide food, water and air. If anything, these designs maximized our separation from living nature, sealing off its essential life support functions in vats and tubes, except where it was pleasing to the eye to have a pretty lawn or flower garden.

By contrast, my work was becoming increasingly focused on our physical, psychological and social need to live fully within nature in all of its wildness, diversity, robustness and efficiency. To call attention to the difference between this mode of living and the other schemes, I eventually coined a word for it:

> **Gaiome** ('gī'ōm) n, an artificial world in space that sustains itself
> using natural ecology. From *Gaia*, the theory of Earth as a living,
> self-regulating organism.

Gaiomes began as a modest attempt to update Tsiolkovsky and O'Neil's designs, which had been gathering dust for decades. But as question after question led back to ecology, I ended up with something unexpected: a direct challenge to the story of escape and conquest that drove space exploration for over a century. Where space colonies once promised endless growth for our current way of life, gaiomes, as living ecosystems, would require us first to find a new way to live. I also began to see how the same pattern of war and waste that threatens us on Earth has measurably begun to cripple us in space.

Make no mistake: human space flight is in jeopardy today. Despite mega-leaps in computer and materials technology since 1969, it has become more costly and dangerous than ever to send people into orbit. Neither money nor new technology nor political will has cracked the problem thus far. Nor, in my view, are they likely to. Working on what we must *do* to make space habitation possible is like asking what a caterpillar must do in order to fly. Wrong question! It can't fly. The important thing to ask is what it can *become*.

Are we the kind of civilization that can live large in the universe? Chapter 1 (*Far and Away*) dissects the space frontier myth to discover that for the moment, we are not. The evidence suggests that we aren't even qualified to live here on Earth, where we have all but spent the abundant inheritance of evolution. In order to survive beyond Earth, the chapter concludes that we must first embrace ecology here on Earth, where the lessons are easier and we have the most help from species that evolved alongside us.

Still, we have become conscious of the wider cosmos; it would be a shame to turn our backs on it. Long before we can become a cosmic species, we will need everyday space travel. Chapter 2 (*Space for Everyone*) borrows engineering and economic models from other transportation industries to establish eight minimal criteria for safe, routine and sustainable passenger space flight. The chapter then proposes a strategy to achieve these goals.

Astronauts today echo our consumer culture by using prepackaged stores of food, water and air lifted at great expense to orbit. Ecologically speaking, this does not even qualify as life support. Chapter 3 (*Gaia and Her Children*) examines how life makes the connections necessary to support itself regeneratively on three scales: globally, in sealed terraria, and in thousands of permaculture farms and gardens world-wide. The chapter will extract from this discussion seven lessons and six heuristics that will help us to design self-supporting worlds.

Long before any settlers depart for lands beyond Earth, their choice of where to live will begin to shape their values. Chapter 4 (*New Worlds, Found and Made*) prospects the solar system for suitable places, revisiting numerous past schemes for settlement and their likely social consequences.

With the foregoing lessons in mind, the next two chapters discuss how we might design and build a living world. Chapter 5 (*Design*) describes the constraints placed on gaiome architecture by the space environment and lays out some traditional and original design solutions. As with Chapter 1, you may glimpse the occasional statistical reflection of your own being: as a consumer, as beneficiary of the vast web of life, as a composite organism. Chapter 6 (*Construction*) looks at who might build these tiny worlds, how they might do it, how soon and at what cost.

As I developed the detailed computer models for these chapters, it became clear that a "master plan" for space travel and habitation would be premature. Too many basic astronomical, biological and political questions remain unanswered or even unasked. Still, I have backed my work with the most accurate data I could find, so that hard engineering numbers can join the hard lessons from ecology as the basis for a new discussion. *Gaiome*, then, is not so much a proposal as a new way to talk about space; a challenge to get out of our present rut and take a fresh look at what it means to be alive in the wider cosmos.

What kind of civilization would build gaiomes? Who would live in them? How would they change with time? Chapter 7 (*Adaptation*) will discuss gaiomic life and its prospects, first on the scale of humanity's near future, then on evolutionary and cosmic time scales. As regenerative ecosystems, gaiomes would qualify as living organisms in their own right, with unique consequences for their residents. These extrapolations illustrate the advantage of biological and cultural diversity, rather than total energy use, as a measure of cosmic progress.

Throughout these pages, you will encounter space not as a frontier for human conquest, but as an evolutionary challenge for all of Earth life—including your own. Chapter 8 (*Homework*) invites you to seize the challenge of cosmic metamorphosis, not through esoteric practices, nor by joining or renouncing any organization, but through deliberate choices in your everyday routines and relationships. These "assignments" outline the work necessary to make a lasting home for ourselves on Earth and beyond.

Measures

Planning anything from backyard gardens to worlds requires measurements and estimates. Given the ease of modern global communications, it is completely inexcusable that space agencies and their contractors continue to use an archaic grab-bag of units of measure. Mars Climate Orbiter crashed, in part, because the contractor used imperial units, while the navigation team used metric. Let's not invite that type of error here.

Numbers in this text appear in short scale, which groups thousands with commas. In this system, a thousand million (1,000,000,000 or 10^9) is one billion. Since many of these numbers come from computer models, a few of them may contain rounding errors in the final digit. Financial figures appear in U.S. dollars, inflated to 2005 values unless otherwise indicated.[6]

For physical measures, I always specify the unit and stick to commonly accepted variants of Système International (SI)—the modern form of the metric system, recognized world-wide. Length appears in meters (m, the SI unit) or sometimes microns (10^{-6} m) or kilometers (1 km = 1,000 m); mass in kilograms (kg) or tonnes (1 t = 1,000 kg); time in seconds (s); areas in square meters or hectares (1 ha = 10,000 square meters or 1/100 of a square kilometer) and temperature in degrees Celsius (°C), which you can convert to Kelvin (K, the SI unit) by adding 273.15? Absolute zero, the minimum possible temperature, is 0 Kelvin.

You will encounter three non-metric units of measure. First, gravity and acceleration appear in gees. One gee is the gravitation we feel on Earth. In a rocket or roller coaster with two gees acceleration, for example, you would weigh twice as much as you do on the ground. Second, it is sometimes convenient to express speeds in Mach numbers, or multiples of the speed of sound (330 m/s). A Boeing 747 airliner can fly as fast as Mach 0.95. Third, the average distance between Earth and the Sun is called an Astronomical Unit (AU). Mars orbits the Sun at an average distance of 1.5 AU, half again as far away as Earth.

If you're unfamiliar with metric units, here are some helpful conversions for later reference: A meter is about 39 inches; a square meter is about 10 square feet; a hectare is about 2.5 acres, or a square the length of a football field on each side. There are 1,609 meters in a mile and one meter per second is about 2.2 miles per hour. A kilogram is 2.2 pounds and a tonne is 2,205 pounds. At sea level, water freezes at 0 °C and boils at 100 °C, while in Fahrenheit, it freezes at 32 °F and boils at 212 °F. A temperature of 300 K is about 80 °F.

Now let's begin.

Chapter 1

Far and Away

"Viewed from the distance of the moon, the astonishing thing about the earth, catching the breath, is that it is alive."
—Lewis Thomas[7]

December 24, 1968: A gray moonscape, framed by the metallic window of the Apollo 8 spacecraft, glides across the screen at a stately pace. A quarter of humanity has crowded around any available TV or radio to witness the broadcast—the first from another world.

For the past five orbits, the astronauts have caught sight of Earth, dazzling blue, rising over the cratered lunar plains. No one has ever seen our home world from so far away. Mesmerized, they snap pictures of it every chance they get. One image in particular stands out. The Earthrise photo will soon become the emblem of its time, appearing on magazine covers, newspaper articles, logos and even postage stamps for decades to come. The radio chirps and the voice of Command Module Pilot James Lovell struggles to bring this mythic sight home to 750 million listeners. "The Earth from here," he tells us, "is a grand oasis in the big vastness of space."[8]

Earth, so far as we know, is unique in the universe. Over the past fifty years, space probes have mapped dozens of Earths worth of new land among the planets and moons that orbit the Sun. Optical telescopes have found hundreds of planets around other stars. Radio telescopes have scanned thousands of stars for intelligent signals. Yet none of these efforts have found even a trace of life. Only here on this world of forest greens, ocean blues, delicate flowers and sunsets, have five to thirty million species made their home.[9]

Home. Does our world need any other name? All who live here are terrestrials, a word derived from the Roman Mother goddess Terra Mater, whom the Greeks called Gaia. Other languages from ancient Egyptian to modern English name the world for its soil. But far from profaning our planet's beauty, this humble name celebrates it. After all, a handful of Earth supports as many as 10,000 microbial species, and they in turn play an important role in supporting us. Though people push parts of it around and often treat it, literally, like dirt, there is no place like Earth, our undeniable home.

Walk awhile in a high desert, a quarry, a cave, a volcanic scarp, and the land's patterns will begin to hint at how they came to be. Spend time reading and walking with skilled geologists (*geo-*, again from *Gaia*), and the land will slowly surrender its secrets like an ancient text. In recent centuries, great libraries have filled up with geological observations, interpretations and debates. Gradually, through numerous lines of evidence, the threads of a planet-wide story vastly older than the human species have emerged and begun to knit together.

The story is incomplete and fragmentary, though, because geology, water, weather and life have reshaped Earth's surface many times over, erasing numerous lines of evidence. But Earth is part of a solar system in a galaxy literally filled with the scattered remnants of its formation and history. Many new clues have come to light in recent years as astronomers have combed the spectra of stars and planets, geochemists have analyzed the composition of meteorites and geologists have pored over 40 years of data from hundreds of planetary probes. Meanwhile, archaeologists and molecular biologists have made enormous strides in understanding the evolution and complexity of life. Together this work, some of it as recent as last month, has snapped the story of Earth into better focus.

Only this vast tale can properly frame the significance of space travel and the brilliant but flawed civilization that invented it.

The Book of Earth

If the history of Earth could be told in one million pages, it would fill 1,000 thick volumes. At twenty volumes per bookshelf and one hundred volumes to a bookcase, the whole set would take up ten bookcases. Even in such a massive history, each page would have to cover a lot of ground: about 4,570 years—

Mya	Page	Event
Hadean Eon		
4,750	1	Protosolar nebula collapses
4,550	4,377	Earth and Theia form
4,433	8,097	Theia collides with Earth; Moon forms
4,470	21,882	Sun turns onto the Main Sequence
4,000	124,727	Late Heavy bombardment
Archaean Eon		
3,800	168,491	First Prokaryotes
3,500	234,136	Photosynthesizing bacteria
3,000	343,435	Aquatic O_2 producers
Proterozoic Eon		
2,500	452,955	Oxygen Catastrophe
2,100	540,482	First Eukaryotes
1,200	737,418	Sex Invented
1,000	781,182	Multicellular algae and seaweeds
Phanerozoic Eon		
565	876,368	Cambrian Radiation
488	889,497	Cambrian-Ordovician Extinction
475	896,062	First plants
400	912,473	First insects (Devonian Period)
220	951,860	First dinosaurs (Triassic Period)
130	971,554	First flowers (Cretaceous Period)
65	985,777	C-T impact; dinosaurs extinct
35	992,342	First grasses (Cenozoic Epoch)
10	997,812	First monkeys (Miocene Epoch)
3	999,334	Australopithecus africanus
0.154	999,967	Mitochondrial Eve (Pleistocene Epoch)
0.100	999,979	First modern humans (Holocene Epoch)
0.074	999,984	Population bottleneck
0.027	999,995	Neanderthals extinct
0.015	999,997	Recent glaciation ends
0.011	999,998	Green Sahara; Dog and pig domesticated
0.008	999,999	Wheat crops; *Writing*
0.003	1,000,000	Iron tools; humans multiply 400-fold

Table 1.1: Selected events in the Book of Earth.

perhaps a decade per word. An abridged contents might read like Table 1.1, which lists times in millions of years ago (Mya). The table has been skewed near the bottom to emphasize human development, but the page count helps to put events back into proper perspective (note where written history begins). The Book's basic plot would go like this:[10,11]

In the first volume, a disturbance in the galaxy compresses part of a dark nebula of dust and gas. Evidence such as the presence of Magnesium-26 in grains found in meteorites suggests that the disturbance was a supernova, a known source of the very short-lived parent isotope Aluminum-26. The self-gravity of the nebula, now suddenly denser, overwhelms the gas pressure and galactic tides that previously prevented its collapse. Nothing challenges the momentum of its slight rotation and, like a figure skater pulling in her arms, the nebula spins faster as it shrinks. The parts with the greatest spin find orbits that resist gravity; the rest fall inward, flattening the nebula into a thin disk.

In the center of the disk, a dense knot of gas is forming a protostar. Glowing brightly from the heat of accretion, but not yet big enough to sustain nuclear fusion in its core, it grows as material rains down onto it from the collapsing nebula.

Grains of dust begin to stick together; some of the surrounding gases freeze to their surfaces, forming tiny motes that will grow to comet-like planetessimals. Gradually, a few of them, in the denser spiral knots of the nebula, get large enough for their gravity to start pulling in objects at a distance. Planets begin to form.

By about the fifth volume, Earth has formed, its volatiles (water, methane, other easily-boiled liquids) hissing violently into space as billions of huge planetessimals crash into it. Already molten from the impacts, Earth's heavier elements such as nickel and iron sink to form a core, converting enormous amounts of gravitational energy into heat. Short-lived radioactive elements decay in the same time frame, keeping the early Earth hot.

Near Earth, possibly in the same orbit but leading or lagging by $60°$, a smaller sibling of sorts has begun to grow.[12] Astronomers have informally dubbed the planet Theia, for the ancient Greek Titan who in myth gave birth to the Moon Goddess.[13] Because Theia is small, her orbit is stable. But she and Earth continue to grow as impacts add to their bulk.

By the ninth volume, Theia, now Mars-sized, has too much gravity to remain in one place. Or perhaps she's jostled out of position by the planetessimals that continue to rain down on her. Whatever the cause, she falls, gradually, toward Earth.

The worlds collide at a glancing angle, greatly increasing Earth's rotation rate. As debris flies everywhere, the iron cores of the two molten worlds merge. Theia is gone. In fewer decades than the words of this sentence, the debris gathers and forms the Moon.

By the 22nd volume (now we're on the second shelf of the first bookcase), the center of the growing proto-Sun has become dense and hot enough to sustain nuclear fusion. The Sun begins to shine with a light that will not go out for over ten billion years—twice the length of our massive history. The Sun is a G2 dwarf, a bright star that will, throughout its life, outshine 95% of the stars in its neighborhood.[14] Its intense light begins to sweep the young solar system free of gas.

By the 101st volume (at the top of bookcase 2), Earth has cooled enough for a solid crust to form and thicken. Between volume 125 and 170 (shelves 2 through 4), planetessimals rain down on the inner solar system, cratering the Moon and bringing volatiles such as water, carbon dioxide and methane to Earth. The cause of this Late Heavy Bombardment remains a mystery. Lunar craters from this event, undisturbed by wind or water, survive to the present day. Earth's craters, by contrast, typically don't even last ten pages.

Well before the end of bookcase 2, life appears on Earth. Because weather, geology and subsequent life have erased so much from this era, we don't know how it got here. The amino acids that form proteins occur naturally over a wide range of environments, including deep space. For all we know, life could have begun off-world and come to Earth as hardy cells entrained in the Late Heavy Bombardment. But this notion, called exogenesis, merely side-steps the central question: how did the simple building blocks of life get together to form cells capable of metabolism, reproduction, adaptation, movement and self-defense? In other words, which came first, the genes (DNA and RNA) that encode the instructions for building the cell, or the cell itself? To date, no one knows the answer, though like the interlinked parts of most living systems, the two may have evolved more or less in parallel.

The earliest life forms in the fossil record are prokaryotes: micron-sized cells with hard walls and no nuclei. Their world, the young Earth, is one of extremes. The atmosphere, mostly made of carbon dioxide, is a hundred times thicker than it is today. Thermal vents perforate the ocean floor and the land oozes with hot springs. These environments host huge variations in temperature, salinity and acidity. As autotrophs (Greek for "self-nutrition"), early cells use these variations to obtain energy, extracting carbon from CO_2 and oxidizing electron-donating substances such as sulfur to form acids.

Mutations occur, most of them fatal. Cells that use energy too quickly burn up or starve. Cells that use it too slowly are crowded out. In their dynamic environment, many cells find themselves challenged by scarcities of nutrients, extremes of temperature and other adverse conditions. Here, the occasional mutation proves advantageous, extending the range of the next generation. Each organism is a little experiment conducted under life-or-death pressure to adapt, repair, conserve or reproduce. Some 10^{30} organisms—quadrillions of quadrillions of them—come to occupy the Earth. [15]

Because prokaryotic cells divide on a time scale of minutes to hours, they can evolve tens of thousands of times faster than do modern mammals. The prokaryotes also exchange genetic information through a variety of means such as bacteriophages and direct contact, allowing communities to share beneficial traits within the same generation—further speeding their evolutionary development. Even so, it takes a full 65 volumes of this type of natural research and development to find a new source of energy: the Sun.

By volume 235, in water supersaturated with salt, a strain of halobacteria develops a new pigment (bacteriorhodopsin) in its cell wall. When struck by solar photons, the pigment flexes, pumping protons out of the cell. The chemical gradient thus generated provides the organism with a potent power source. This method survives to the present day as the only mode of proton-transport photosynthesis on the planet.

Electron transport photosynthesis as seen in plants does not appear for more than one hundred more volumes. By volume 350 (fourth bookcase), cyanobacteria, commonly called blue-green algae, have begun to use a pigment called chlorophyll to capture the energy in sunlight. In the somewhat complicated Calvin cycle of chemical reactions, cells store and use this energy to pull carbon out of the CO_2 atmosphere, producing oxygen as a waste gas.

This method of gathering energy becomes so successful that it begins to enrich the oceans with oxygen that, in turn, binds with dissolved iron to produce banded iron formations on the sea floor. But by about volume 450 (bookcase 5), oxygen production has overwhelmed the ocean and come to dominate the atmosphere. Most bacteria cannot survive this oxygen catastrophe, so the majority of Earth life retreats to anoxic environments—clays, deep ocean sediments and the like. Though the world has become harsh for them, the anearobes do not die off, but survive to find vital roles in the exotic ecologies to come. For example, with the emergence of plants 1.6 billion years later, some anaerobes will become nitrogen fixers in plant root zones. Thus even very primitive life proves its tenacity, adaptability, and ecological potential.

Not all life retreats, however. By the time of the oxygen catastrophe, the prokaryotes have undergone ten trillion generations of evolution comprising perhaps 10^{43} individual experiments.[16] Some cells have started to adapt to the new chemical environment, developing the citric acid cycle to utilize oxygen for energy. As these new aerobic organisms evolve, not only does genetic material continue to flow between them, but smaller cells sometimes find their way entirely inside some of their larger neighbors. Somewhere between volumes 540 and 650 (bookcases 6 and 7), a few of these tiny endosymbiotes evolve to become organelles ("little organs") that provide various services within their host cells. Among the many types of organelles are chloroplasts that use sunlight to produce Adenosine triphosphate (ATP: life's energy currency), mitochondria that put the ATP to work, and nuclei that protect the DNA within an inner membrane. Together these form a new domain of life, the eukaryota: strange composite beings-within-beings (the mitochondria retain their own DNA) that will eventually evolve to include all protists, fungi, plants and animals.

With the advent of aerobic eukaryota comes the dawn of the solar economy. From this point forward, nearly all natural wealth will derive from sunlight and the autotrophs (such as plants and algae) that harness it. Eukaryotes not blessed with chloroplasts are heterotrophs: they must get their nutrition by consuming other organisms.

By volume 740, as Rodinia, the first known supercontinent, begins to form along the equator, life has been passing genetic information to peers and offspring for 2.6 billion years. Then, over the course of some 40 volumes, life discovers sex, a mode of reproduction that significantly increases genetic variation among offspring. Although it has its costs (a sexual organism may not be able to reproduce when stranded in a new environment), sex greatly increases the rate of variations so essential to the process of evolution.

In volume 782, metazoans (multicellular organisms) such as algae and seaweeds start to appear. Rodinia breaks up by volume 836. Sponges, jellyfishes and flat worms appear by volume 870, and the latter two evolve simple nerve cells. Seven volumes later, life begins a nine-volume surge in size, complexity and diversity known as the Cambrian radiation. By the end of this relatively short period, the world is filled with oxygen-breathing animals with nervous systems. Some of them venture onto land.

Life moves in fits and starts. Ten volumes before the end of the ninth bookcase, climate change kills a significant fraction of the world's species. Perhaps it was glaciation; perhaps depletion of oxygen in the ocean. Whatever the cause, it triggers the first major mass extinction event. To qualify as "major," more than 30% of the genera (groups of related species) must die off. Life will confront five more major extinctions before the end of our story, taking, in each case, millions of years to recover its former diversity.[17]

The first land plants appear by volume 897, and the first insects and sharks appear by volume 913. We're now in the tenth and final bookcase. Plants evolve seeds in volume 922, improving their ability to spread into new terrain. Reptiles appear in volume 935; dinosaurs in volume 952. Plants develop flowers by volume 972, co-evolving with their pollinators (mainly insects) to produce an enormous range of new species.

In volume 985—on the bottom shelf—an asteroid impacts the Earth, wiping out half of all species, including the large dinosaurs (a few of the smaller ones survive and eventually evolve into birds). This is the fifth major extinction. Mammals spread and diversify. Grasses appear in volume 993.

By the beginning of volume 1,000, the current book, early human ancestors have evolved away from the line that will become chimpanzees and bonobos and begun to walk upright. Proto-humans become completely bipedal with the emergence of Australopithecus in Kenya around page 344. By page 600, Homo Erectus is one of several proto-human lines that make their appearance in East Africa and, over the next few hundred pages, spread into Europe, Western Asia and Australasia, evolving adaptations to the local conditions.

Like all animals, humans inherit their mitochondria from the ovum. Geneological studies of mitochondrial DNA in people from every continent place the most recent female ancestor of all modern humans—Mitochondrial Eve—at around page 967.

Isolated groups evolve rapidly and become distinct species. Homo Neanderthalensis, adapted to the cold, comes to live in Europe by page 972. The first modern humans, *Homo Sapiens*, appear in central Africa by page 979.

On page 984, the human population suddenly contracts, perhaps to only a few thousand individuals. The population bottleneck, as it's called, coincides with the Toba supervolcano eruption in Indonesia. Three thousand times more powerful than Mt. Saint Helens, the explosion and subsequent ash plume may very well have darkened skies over much of Earth, reducing global temperatures by $3\,^{\circ}$C or more. While no big deal for life in the long run, this climate change spells disaster for itinerant human tribes struggling to survive in unfamiliar environments. Long winters and late blooms wipe out essentially all human settlements beyond central Africa.

Gradually, the population grows back. A new set of tribes begins to explore, reaching coastal Europe and Australasia by page 990 (around the time of Y-chromosomal Adam, our most recent male ancestor). By page 994, they have reached Japan, the Bering bridge and North America. The last known Neanderthals die on page 995, around the time that some of the human tribes learn weaving. A long glaciation ends around page 997, and sea levels rise, covering up the Bering bridge to North America and obliterating much of the fossil record along all coasts. Large mammals go extinct in prodigious numbers.[18,19]

By page 998, human migrations finally end at Tierra del Fuego, at the Southern tip of South America. Our global population reaches 5 million. Some tribes domesticate dogs; some set about domesticating plants. Agriculture shifts diets toward grains, and the energy thereby gained increases fertility. Human population grows, and with it, the demand for more arable land. On page 999, the first domesticated cereal, wheat, appears in the fertile crescent and domesticated food animals such as pigs begin to graze the green savanna of the Sahara. Plowing and weeding decrease biodiversity over large tracts of land, accelerating soil erosion and nutrient loss. The storage of grain blunts otherwise rapid ecological feedbacks, leading to land mismanagement. Vast tracts go barren, forcing migrations and expansion. Civilization appears and begins to keep records in ever more detailed and abstract systems of writing. By the end of the page, there are perhaps 15 million humans.

On the very last page, human intelligence seemingly triumphs. The gradual accumulation of knowledge beats back death in small increments; our population rises, doubling at first every millennium, then every century. Hebrew and Greek alphabets detach writing completely from pictographic depictions of the world, leading to a revolution in abstraction. Arabic numerals (including the all-important zero) enormously simplify computation. In the final, short paragraph, coal and oil massively amplify our power to move about and reshape the world. In fewer words than this sentence, ships span the globe, engines reshape the land, planes conquer the air, computers catalog the genome and space probes map dozens of new worlds. Our population has soared 400-fold on a single page, and doubles now every four to six words.[19,20]

In this sprawling epic, civilization occupies just two pages out of a million. A human life lasts just a few words. Yet somehow, we've arrived at the story's most dramatic moment. We can see it in the Earthrise photo: the sharp boundary between the endless void and life's blue sphere; the unseen photographer who has climbed beyond life's circle of gravity; a dawning sense of place in the wider cosmos; a moment pregnant with promise.

A Challenge to Evolve

Four decades before the Earthrise photo, a fourteen year-old boy glimpsed something of its promise. In an article published in the February 15, 1927 issue of *Deutche Jugendzeitung* (Journal for the German Youth), he wrote: "An age-old dream of mankind—to travel to the stars—appears to approach fulfillment."[21] The opportunity was unprecedented: a giant leap. With effort and good fortune, perhaps he could bring it into being.

The student's name was Wernher von Braun, the man who eventually would design the Saturn V rocket that carried the Apollo astronauts to the Moon. But he was writing long before the first satellites, indeed in a time when rockets were merely dangerous toys.

Three years later, von Braun joined the VfR (*Verein für Raumschiffahrt*: "Space Flight Society"), a 900-member amateur rocket club.[22] Perhaps its most prominent member was Hermann Oberth, whose self-published 1923 monograph *Die Rakete zu den Planetenräumen* (By Rocket into Planetary Space), had inspired von Braun and created an international sensation.

Learning of Oberth's work, the pugilistic Soviet government quickly announced that a Russian had long since invented everything in his book and more.[23] Oberth investigated. It turned out that a deaf Russian school teacher, one Konstantin Eduardovich Tsiolkovsky by name, had published hundreds of technical monographs and science fiction pieces about space travel starting as early as 1895. Not only had Tsiolkovsky long-since invented the fundamental equations of rocketry, he had, with uncanny detail and accuracy, anticipated many of the problems and sensations of modern space flight.[24] Oberth was so impressed that he became a lifelong fan of Tsiolkovsky and promoted his work abroad.

Tsiolkovsky's sweeping cosmic vision began with the Sun, which he recognized as life's primary energy source. Because Earth intercepted only one part in 2.2 billion of the Sun's total light output, life could only achieve a tiny fraction of its potential from the surface of the home planet. Tsiolkovsky reasoned, with an almost Marxian sense of inevitability, that humans would eventually "leave the cradle" of Earth. Only then could we truly make something of ourselves, using more energy than is available on Earth to live large in the universe.

To Tsiolkovsky, space was nothing short of a new evolutionary challenge. Beyond Earth lay lands so vast and varied that all of life's journeys by ocean, land and air would pale by comparison. In his eyes, space travel was the grandest step evolution would ever take because it would unchain Earth life, after billions of years of confinement, to achieve endless destinies.

Tsiolkovsky then went on to show that rockets were the means to this end. This was where von Braun and his VfR colleagues would make their mark. Their goal, always, was interplanetary travel.

Their budget, however, was tiny, and so were their rockets. These they launched from a Berlin suburb, charging gawkers admission to defray costs.

Von Braun's fortunes changed in 1932, however, when the German army invited him and other VfR members to develop rocket weapons. Von Braun gladly accepted, using the army test range for his experiments, skipping the second half of his undergraduate course work and completing a doctoral dissertation on rocketry in a scant 18 months.[25] Others in the VfR dragged their feet, suspicious of an increasingly violent and paranoid regime. Two years later, after the Nazis had consolidated their power under Hitler, financial difficulties and ever-tighter regulation forced the VfR to disband. Some members, such as Willy Ley (who would become a prominent American author), soon left Germany altogether.

Rockets interested the Nazis because, as it happened, the Treaty of Versailles at the end of World War I prevented Germany from developing long-range artillery. But the treaty said nothing about rockets. So, in a strange twist of fate, the brutally repressive Nazi government became the first to sponsor the space dreamers.[21]

Thus a young Wernher von Braun found himself on the fast track as a Nazi, eventually joining the SS (*Schutzstaffel*: "protective squad"), a paramilitary organization that chose its members for their supposed racial and ideological purity. Before World War II was through, von Braun had attained the high rank of Major in the SS, overseeing a secret island factory called Peenemünde. Here, thousands of slave laborers from the nearby Mittelbau-Dora concentration camp built the rockets in a labyrinth of caves carved into a mountain. Despite aerial bombing, wartime supply problems and active sabotage by the laborers, von Braun oversaw the building of over a thousand V-2 rockets (V for *Vergeltungswaffe* or "Retaliation weapon"). Near the end of the war, the missiles rained down on England, killing over 2,700 civilians.[26] The greatest damage, however, occurred in Germany, where as many as 15,500 prisoners may have died building the rockets.[27]

Designed By a Nazi

When the war ended, von Braun and hundreds of other former Nazi rocketeers sought out and surrendered to the American army as it overran Germany. The U.S. Army seized Peenemünde, shipped all the V-2 parts it could back to the states and left the island and its production staff to the invading British and Soviet armies. The United States put the captured Germans to work on rockets at the Army's Proving Grounds in White Sands, New Mexico and Fort Bliss, Texas.[23][28]

Von Braun had been a quick study of the Nazi path to power, especially its manipulation of rich media images and spin control. As soon as anti-Nazi sentiment had cooled off somewhat in America, his dreams figured prominently in a series of lavishly illustrated articles in Collier's Magazine.

Here was his plan to conquer space: (1) Build huge rockets. Control them by radio from the ground. (2) Perfect the technology needed to get them into Earth orbit. (3) Put life-support systems in them and send monkeys, then men. (4) Build a re-usable rocket, or Space Shuttle, and make manned missions routine. (5) Build a giant, permanently-manned space station to study weather, relay radio signals, conduct experiments, spy on enemies, etc. (6) Build a permanent military base on the Moon. (7) Use the space station as a base to assemble a giant fleet of nuclear-powered space ships to send hundreds of military men to explore Mars.[29]

NASA adopted his agenda wholesale when it hired him to beat the Soviet Union in the race to the Moon. The U.S. Government and media turned a blind eye to the German rocket team's Nazi past, installing von Braun as Director of Marshall Space Flight Center and his associate Arthur Rudolph, who would later renounce his U.S. citizenship,[30] as program manager for the Saturn V rocket.

Standing 111 meters tall (seven times the height of the V-2), with a lift-off mass of 2.8 million kilograms, the Saturn V was the mightiest rocket ever built. Of the 13 launched, not one ever failed. On July 20, 1969, it achieved the goal of putting Americans on the Moon, winning the "space race" once and for all.

With its primary goal achieved, NASA continued forward for a time on raw momentum. NASA's contractors now had the ability to crank out several Saturn V rockets per year, but only a handful more would be launched. On December 14, 1972, the last of six missions left the Moon. Money and political will had run out.

But NASA refused to die. By then, it had status, infrastructure and a cadre of super-educated dreamers with some unfinished business: items 4-7 of the von Braun playbook. And so, as political support waned and budgets dwindled, NASA got busy promoting the Space Shuttle, the Space Station, and planning permanent Moon and Mars bases.

Through much of its history, NASA's annual budget was decided in the U.S. House and Senate Appropriations subcommittees on the Veteran's Administration, Housing and Urban Development, and Independent Agencies. (The VA-HUD committees were disbanded on March 2, 2005, and NASA was reassigned to the Commerce, Justice and Science subcommittee.) Because its budget historically was carved out of appropriations that might otherwise house the poor or care for wounded veterans, the social relevance of NASA's activities fell under especially intense scrutiny.

How to convince the skeptics? Cold War military necessity could no longer quite hold the day. What sold better was the frontier.

The Final Frontier

The American frontier had closed, of course, long before the dawn of space flight. The United States Census of 1890 noted its passing, remarking that the unsettled areas of the country, once large tracts that lay beyond advancing lines of civilization, had at last broken up into mere pockets of wilderness. Historian Frederick Jackson Turner, speaking on "The Significance of the Frontier in American History" at the Chicago World's Fair in 1893, advanced the view that the frontier had been the central defining feature of the United States and its citizens.[31]

Turner framed his Frontier Thesis in evolutionary terms: pioneers would regress nearly to "savagery" in order to survive the unsettled wilderness. Then, as civilization filled in behind them, it would progress steadily through ever more advanced industries toward modernity.

To Turner, the frontier explained not only America's rugged individualism, but also its paradoxically strong national identity. He wrote:

> "Nothing works for nationalism like intercourse within the nation. Mobility of population is death to localism, and the western frontier worked irresistibly in unsettling population...What the Mediterranean Sea was to the Greeks, breaking the bond of custom, offering new experiences, calling out new institutions and activities, that, and more, the ever retreating frontier has been to the United States...and with its going has closed the first period of American history."

From NASA's earliest days, Turner's paper circulated widely throughout the Agency.[32] Administrators and engineers alike readily perceived the Moon as the threshold to an infinite frontier that not only could recapture the central motive force of American history, but magnify it without limit.

American science fiction throughout the 20th century was rife with visions of a space frontier. Pulp fiction and movies filled the young minds of a generation with whole constellations of images: Astronauts bouncing buoyantly about in space suits. Rooms full of mission controllers with eyes riveted to giant status screens. A rocket pilot gripping the controls as the clock ticks backward toward ignition. The Earth looming large in a view port. Diners squeezing supper from plastic tubes. The incessant clatter of fans, pumps and actuators providing life support. Maverick tycoons mining the Moon for profit. Militaries planting flags and racing to control the "high ground" and orbital "chokepoints."[33] Rockets running gauntlets of extreme cold, isolation, orbital debris, solar flares and radiation belts to ply new trade routes between gigantic space stations and bubble-domed colonies in Lunar craters. Meanwhile, in nearby crevasses under black, star-speckled skies, growing mounds of trash neither rot nor rust away in gray wilds so still that a footprint there might last a million years.

Beyond its air-brushed glamor, adventure and novelty, the space frontier made three more promises: endless growth, escape from tyrants and disasters, and transcendence of the physical limitations of Earth and flesh. To Cold War America, these notions sold like instant hotcake mix. But were they accurate?

Growth

Increasing one's personal wealth is the central idea of modern civilization. To speak out against it is to be branded anti-progress if not anti-American. If you expect to have more than your parents, and your children to have more than you, then you are deeply invested in the concept. Investment itself implies growth.

The pattern of growth we're talking about occurs often enough in nature. Whatever increases by a fixed fraction of itself in a fixed amount of time will grow exponentially. At the moment you were conceived, you were a single cell. Within 18–20 hours, that cell divided in two. Over the next 18–20 hours, those cells divided again, and so on. Early in your mother's pregnancy, you grew exponentially.[34]

Populations grow exponentially when the birth rate exceeds the death rate. Not only is human population growing exponentially right now, but the rate of increase itself has been increasing (with only very short-term exceptions), for thousands of years.

The world economy is also structured around the assumption of continued growth. If your money is earning interest, growth is working for you: your nest egg will double in a finite amount of time.

The doubling time for anything that grows is about equal to 70 time units (such as years) divided by its percentage rate of increase (such as an annual interest rate). Human population has doubled in the 42 years since 1965, so the average annual growth rate over that period is $70 \div 42 = 1.67\%$. The growth rate varies somewhat from year to year and is currently about 1.17%. This would give a doubling time of $70 \div 1.17 = 59$ years.[20]

Growth would seem an absolute social good. For example, in the 20th century, the population of the United States grew by a factor of 3.5.[35] Not co-incidentally, over this time period, the rate of copyright registrations increased fivefold; patent registrations increased sixfold.[36-38] Many of these filings involve media (such as film and digital works) and physical phenomena (such as quantum mechanics and DNA) that were unknown in 1900. If civilization grew by a factor of ten, it seems fair to expect the sort of discoveries we now see every decade to emerge every year—almost as soon as new needs are identified. Now imagine what could happen if civilization grew by a factor of 1,000. Could it grow forever?

Not on Earth. The world has only so much land, and we depend on it for a lot more than housing and transportation. All of our food and most of the energy we use comes from the solar economy—sunlight converted by plants into the primary sugars that power the biosphere. Even fossil fuels—oil and coal—are stored (and substantially degraded) energy from a half-billion years of buried plant life.

Sunlight delivers a steady 1.8×10^{17} watts of power to Earth. Plants convert about a thousandth of that into biologically useful forms such as carbohydrates.[39] In 2002, human energy consumption came to 7.6% of the energy that plants stored that year, and half of that went to the richest billion of us.[40,41] If the remaining 5.5 billion rose to the same level of affluence (as China and India are working very hard to do), humans would consume 26% of Earth's total plant productivity. But this figure neglects the effects of habitat loss due to agriculture, construction and pollution. Even in 1992, when there were 1.2 billion fewer of us, Vice President Al Gore suggested that our total consumption had already reached 40%.[42] Ecologist H. T. Odum has pointed out that in stable ecosystems, no single animal species consumes more than 2% of the total plant productivity.[43] Clearly, our energy use alone puts us way beyond the point of natural balance with our environment.

To fill the energy gap, the global economy depends on oil, coal and uranium reserves. These may suffice for a time, but population continues to grow. Unlike an embryo, we have no built-in stopping mechanism short of exhausting our fossil and biological reserves like an overdrawn bank account. Then, as Thomas Malthus famously warned in 1798, war, famine, plague and pestilence would surely follow.[44]

But wait! This dreary picture ignores the existence of lands beyond Earth. The Moon and Mars combined have more land than Earth. The asteroids are rich in resources, from water to soil minerals to precious metals. For over a century, those who grasped the enormous bounty of space have wondered how anyone could believe in limits to growth. All the oil and coal on Earth is stored solar energy, yet as Tsiolkovsky showed, the Sun provides 2.2 billion times as much as reaches Earth. So why bother to save energy? The total amount of steel of all grades in use throughout the world would fit in a ball less than 1,800 meters across.[45,46] The asteroid belt has thousands of pure metal asteroids larger than this,[11,47] so why recycle? The galaxy has hundreds of billions of stars and probably trillions of planets. Beyond that, the best cosmological evidence suggests that the universe goes on forever. When would we ever run out of new material? Why not simply travel beyond Earth and clear new lands as we need them?

Suppose we could. By exporting all excess population to the Moon and Mars, we could avoid further crowding on Earth for a time. At an annual growth rate of 2% (which is low for an open frontier), the new worlds would become as crowded as Earth in 35 years. From that point forward, we would need more land still. By then, there would be twice as many people having babies, so in another 35 years we would need not one new Earth, but two, for a total of four.

Fortunately, the known asteroids contain enough material to build thousands of Earths worth of new land in the form of O'Neill colonies. We are only just beginning to discover the minor planets of the outer solar system, including Jupiter's Trojan asteroids and the Kuiper Belt, which includes Pluto and thousands of other small worlds. Together, these bodies multiply the available material by another factor of a thousand or more. Thus, at the same dry land population density as on today's Earth, the solar system could hold something like 10^{17} people.

At a 2% growth rate, it would take only 870 years, a fifth of a page in the Book of Earth, to fill up every piece of asteroidal, Trojan and Kuiper Belt real estate in the solar system. Even so, this falls far short of using the Sun's total output. To capture all of the Sun's energy, we would need to dismantle major planets such as Neptune or even Jupiter to get enough raw building materials. By capturing 2.2 billion times as much sunlight as Earth and taking care to maximize green spaces, our population may be able to grow to 10^{20} before again claiming an unsustainable fraction of plant productivity. Even so, growing at 2% per year, it would take us only 1,200 years to reach this enormous population. To continue our growth beyond that point, we would need to find land beyond the solar system.

Scale:	Moon/Mars	Solar System	Galaxy
Population:	6.5×10^9	10^{20}	10^{30}
Doublings:	1	34	67
Growth/year	Years to reach Earth's present density		
2%	35	1,200	2,300
1%	70	2,400	4,700
0.1%	690	23,000	47,000

Table 1.2: Growth and overpopulation in the galaxy. The top section shows the populations and the number of doublings needed to match Earth's present dry land density on three cosmic scales. The bottom section shows the time it would take to reach each of these population levels at three different growth rates.

Of course, we could prolong the process by slowing our growth. However, doing so has proven extremely challenging even for strong central governments such as China. Across an open frontier, it's hard to see how even this modicum of restraint could be enforced. But even if growth does slow, as long as it remains above zero, our need for new land will rise exponentially.

Table 1.2 shows the total human populations needed to fill the Moon and Mars, the solar system and the Galaxy to the same density that we have on Earth today: 44 people per square kilometer of dry land. It also shows the approximate number of times our present population would have to double to reach that figure. Finally, the lower part of the table displays the number of years it would take to reach each population level, assuming various growth rates.

The figures are very rough, but this hardly matters. Even if my assumption that the Galaxy contains ten billion empty, habitable solar systems is off by a factor of 100, that's only a difference of seven doubling times (out of 67). The point remains that the Galaxy would fill up in a hurry.

Table 1.2 ignores travel times. Even moving at nearly the speed of light, it would take settlers 100,000 years to cross the galaxy, far longer than any filling time in the right-hand column of the table. The final limit to growth, then, is raw distance.

It may at first seem that the special theory of relativity provides a loophole: viewed from a spacecraft moving at very close to light speed, the galaxy would appear foreshortened along the direction of travel. Thus travelers could cross it in perhaps only a few years of subjective time.

In practice, though, outrunning population growth by relativistic expansion would require the great bulk of humanity to remain in transit between the stars at any given time. As soon as a band of travelers slowed to settle a solar system, their reference frame would shift. The stars would return to their former imposing separations. From the perspective of any solar system in the galaxy, those still in transit would take 100,000 years to cross the Milky Way, their lives prolonged by time dilation. Even at a miniscule annual growth rate of 0.1%, solar systems would fill in 23,000 years, before travelers could get even a quarter of the way across the galaxy. Then populations would double, if unchecked, every 690 years. Thus even with relativistic expansion, all habitable solar systems within the volume of settlement would fill at an exponential rate. Meanwhile, the surface of this volume could not expand faster than the speed of light. No matter how small its rate, exponential growth will overtake a volumetric expansion in a finite amount of time.

Perhaps we could get around this problem by somehow accelerating all new life to ever higher relativistic speeds. Our ever-faster progeny would still need power and materials for growth, of course, but they could never slow down to obtain them. Thus they would be forced (if they were able) to disassemble planets and stars for power and material without breaking stride. Fleeing outward from the solar system in all directions, their lines of communications would stretch ever thinner. Theirs would become a path of total destruction through a universe that speeds by in a blur. They would never know a moment's rest as they rush outward into deeper oblivion.

Long before growth reaches its limits, though, life gets hectic. A competitive economy with obligatory growth cannot avoid suffering. Because the wealth of individuals and regions changes at wildly different rates, a few get much richer faster, while most, especially agrarians, can't keep up with inflation. The wealthiest investor, Warren Buffet, has earned up to 40% a year for decades on end—doubling his wealth every fourteen months. Meanwhile, a third world peasant whose ancestors got by for millennia in the local ecology now may make two dollars a day or less. While that amount may have been fine a hundred years ago, today it is not even enough to open and pay the fees on an interest-earning account. With her government pressing her to grow exotic cash crops such as coffee rather than subsistence foods long adapted to the local ecology, her ability to eat becomes vulnerable to price and currency fluctuations on a global level. As a result, despite the world-wide cultivation of almost twice as much grain as needed to feed the human population, more than eleven million children starve to death every year.[48]

For the many who struggle somewhere between poverty and wealth, the relentless pressure toward ever-greater visible prosperity increases competition, toil and hoarding to the point where these traits dominate. As resources and time dwindle, the incentive to try and rig the game in one's own favor becomes overwhelming. From the poorest sweatshop to the richest multinational, growth begets greed; greed begets corruption; corruption breeds violence.

There is no outrunning exponential growth. It will buck up against hard limits of resources and energy in finite time, no matter how boldly we go about extracting resources from the cosmos. As surely as tomorrow, the day will come when we cannot expect more wealth, power or children than yesterday. Conquering the space frontier would, at most, forestall that day, magnifying the cost to the majority of people who will be losers in the competitive economy along the way.

Escape

The frontier's second promise is escape from a world already feeling the pinch of growth. Today, a fifth of Earth's human population produces more carbon dioxide than plants can scrub from the atmosphere, and the remaining billions are striving to achieve a similar level of industry. Pollution, disease, guns and drugs cross borders continuously, and nuclear weapons continue to proliferate.

For those of us now richer in cash and energy than ever before, it can be hard to appreciate the enormity of the situation. The 1974 *Limits to Growth* study warned that, like an heir squandering his wealth, opulence has blinded us to our predicament.[49] During the oil gluts of the next two decades, the book was roundly criticized as anti-progress, but the story it told was as old as the hills. In *Collapse* (2005), biogeographer Jared Diamond showed that throughout history, societies that failed catastrophically generally did so within a decade or two of their highest levels of population and prosperity.[50] Through multiple crises of negligence, collapse catches civilizations by surprise. While a society's coping mechanisms (from royal tribute to free markets) may be sufficient to overcome any one crisis (such as pollution), the problem with growth is that it depletes all budgets simultaneously. Thus the society must face famine, energy and water shortages, loss of soil fertility and social turmoil all at the same time.

In the past, though, we lived in a world of many civilizations with separate economies. While some collapsed, others lived on. But now that world economies are knitting together into a single, interdependent whole, we live in deeply precarious times.

All the more reason, say spacers such as the theoretical physicist Stephen Hawking, to establish off-Earth colonies now. Hawking has been making the news lately by urging the construction of space colonies before something (such as a genetically engineered virus) has a chance to wipe us out.[51] He's in good company: Gerard O'Neill predicted wars and dark ages brought on by energy and resource starvation.[1] Astronomer Carl Sagan raised the specter of an asteroid impact such as the one that killed the dinosaurs, or an equally devastating nuclear winter.[52] Robert Zubrin warned against the rise of tyrannical world governments.[53] In each case, the authors felt that we had only the narrowest window of opportunity to get into space and dodge the apocalypse. Science fiction author David Brin has called this the "threshold effect."[54] Even if our problems do fundamentally boil down to growth, they argue, going to space now would at least buy us time and resources while we hash out more durable solutions.

The idea is gaining traction. For example, at www.gaiaselene.com (slogan: "...saving the earth by colonizing the moon"), you will find dire warnings about global warming, melting ice caps and our rate of fossil fuel use ("millions of times faster" than nature produces them). The splash page claims that "By 2050 we will need three times as much power and it will have to be three times as clean." They're probably right. But what to do about it? According to their site, small-scale renewable energy production such as solar power and wind energy are not reliable and scalable enough to sustain any significant growth. So Gaia Selene proposes to build large solar arrays on the Moon's poles and beam the power back to Earth. If or when nuclear fusion becomes viable, they also advocate sifting the lunar regolith (soil) for Helium-3, a relatively non-polluting fusion fuel deposited by the solar wind. There's even a video featuring, among others, Dr. Alan Binder, designer and Principal Investigator of NASA's Lunar Prospector mission, which found evidence for water ice in the Moon's polar regions. According to Binder, permanent, self-sustaining lunar bases pose no problem: where there's water, soils and sunlight, we can grow food. In Binder's words: "What could be easier?"

There's also the Lifeboat Foundation (lifeboat.com), a policy group concerned with limiting and defending against hazards posed by biotechnology, robotics, nanotechnology, and nuclear weapons. Lifeboat opened its doors online a few years ago with detailed renderings of a self-sustaining space station called Space Ark 1 on its splash page. The design consists of four von Braunian space wheels (each 300 meters in diameter) connected in parallel pairs at the ends of a cylindrical bridge. Power would come from solar panels and four fission generators—one in each hub. Space Ark is designed to support 1,000 permanent residents and 500 visitors. Its stated purpose is to provide a backup genetic stock of humanity "just in case."

As famous scientists have joined Lifeboat's board of directors, Space Ark 1 has left the Foundation's main page and buried itself among the links in the site's FAQ. The text that accompanies the many images briefly outlines the Ark's systems: Air (2/3 sea-level pressure and 60% oxygen), Design/Construction, Gravity, Heat, Location, etc. But the only column space devoted to biological life support systems is a brief mention of successful plant growth experiments aboard the late Russian Mir space station.

Is it really such a short step from a few zero-gravity sprouts to ecological life support? Most of us in the space community seem to think so. The prevailing view regards life as so much biomechanical gadgetry, easily and long-since mastered by the disciplines of agriculture and chemistry. O'Neill's colonies, for example, would place farms in sealed, glass-topped tanks outside the colony. By growing each crop in sterile soil, sealed off from the others and from the main habitat, their growth and nutrient needs could be monitored with perfect control and no risk of loss to pests or disease. Yields would be utterly predictable and regular, eliminating any chance of famine.

If only it were so easy. The crops in such a system would need air, water and fertilizer in quantities far too great to haul up from Earth. Thus it would be convenient to obtain them somehow from the habitat's wastes. The habitat itself would need clean air and water, nutritious food and sanitary waste disposal. Food crops by themselves aren't especially well-adapted to purifying raw sewage, so additional machinery would be needed to process materials flowing each direction. To date, Earth-side experiments (which we'll discuss in Chapter 3) have required an enormous amount of machinery and electrical power to provide these additional services, with only mixed success.

Engineering, Life and Life Support

Sooner or later, most science and engineering students realize that their tools, from resistor codes to mathematical formulas, are made to be as easy to use as possible: standardized, linearized, simplified, color-coded, cataloged and explained in every conceivable way in their text books and training manuals. Nature may at first appear to have something in common with human engineering: DNA uses four standard bases; organisms seem to have clear roles in ecologies, such as predator and prey. But this veneer of simplicity is really no thicker than our own perceptions.

The machinery of the living world operates on a molecular scale, so it would seem to be comprehensible through the tools of chemistry and thermodynamics. Yet unlike simple chemistry experiments, life systems down to the smallest organelle are characterized by an astonishing number of connections, many of them quite subtle. Unlike our tools, organisms exist for no external purpose except to live. And to do that, they have evolved what I would consider a defining characteristic of life: the ability to make new connections as needed. This creative aspect of life, coupled with reproduction and genetic variation, allows evolution to optimize organisms for energy efficiency, miniaturization, self-repair, ever-shifting niche roles and countless other traits. Life's technology is literally billions of years ahead of anything humans have invented. Its complexity and chemical interoperability came about through more than 10^{43} experiments optimized across millions of times more parameters and connections than the human mind can hold. [16]

If you asked a group of scientists from 250 years ago to reverse-engineer a cell phone, they would have no prayer of doing so. Even if they could somehow deduce from it the underlying principles of electricity and quantum mechanics, they would lack the manufacturing infrastructure to do anything with it. We are at millions of times less advantage when it comes to making sense of the living world. The only thing that should give us any hope of comprehension is that we, too, evolved as part of it. Thus instinct, intuition and tradition often rightfully guide our research trajectories.

Life abounds on this world because trillions of generations of every organism's ancestors (including our own) participated in the evolutionary process of sorting out and filling niches. This process provides all living beings on Earth with food and living space. The networks that emerge to accomplish this, and all the beings within them, together comprise complex and tightly interlinked ecosystems. Optimizing equations or electronic circuits is not at all the same kind of problem.

Not only does life not recognize intent (except to live), it does not recognize boundaries. Coral, for example, is a composite organism with animal, mineral and plant components. We may identify plants in isolation, but they did not evolve that way. For example, plants scrub carbon dioxide and provide food and shelter to numerous animals at various stages of life. Flowers and seeds co-evolved with birds, insects and their changing environment. Some plants such as sedges and hyacinths can even act in concert with microbes and fungi to purify raw sewage. Putting just one or two plant functions in a black box such as a space farm puts a plant out of several of its other customary jobs—and reassigns them to us. Soon enough, we discover that we are under-qualified. We do not understand life support. No species does. That hard-won knowledge is spread among the many thousands of species that comprise each of Earth's *biomes*, or climate-adapted ecologies.

From deep oceans to marshes to alpine forests, biomes are complex, detailed, interconnected, creative, non-linear and resistant to partitioning. This makes them poor candidates for the tools of engineering, which depend on linearity, predictability, simplicity and modularity. If, in the aggregate, we are not in a sustainable relationship with living ecosystems now, there is no definitive way to tell how much effort it will take to make them sustain us in space.

Indeed, the latest Earth escape club, the Alliance to Rescue Civilization (ARC; www.arc-space.org), acknowledges that any such project would be "very long-term."

Space lacks air, water, food and the ecosystems that regenerate and purify these essentials for us on Earth. We will never have a better opportunity to learn how to sustain ourselves through natural ecology than we have here and now on the home planet.

Since the dawn of civilization, our dreams of growth have put us at war with nature. The easy victories have ended; it's now a losing battle. Irrigated monocrop agriculture sows deserts, bacteria have evolved resistance to antibiotics, and if we don't change course now, up to half of all species will go extinct by mid-century—life's sixth and fastest major extinction.[17,55-57] Given the destruction we've wrought on Earth, our odds of preserving life meaningfully in a radioactive vacuum on a first attempt are indeed vanishingly small. We have shown life so little regard for so long that we never learned the skills of life support. If we wreck the Earth, space offers no escape.

The Myth of Away

If building an ark is off the table, could there at least be a pragmatic middle ground? Could space tide us over with some valuable resources while we learn to live more sustainably? This seductive argument, often heard in space circles, smacks of Augustine of Hippo's ancient prayer: "grant me chastity and continence, but not yet."[58]

Even now, after sixteen centuries, Augustine's pious quip provokes a smirk: a virtue deferred is clearly no virtue at all. If it were convenient to mine space for immediate consumption, doing so would not so much buy us time as postpone the ecological virtues. Space exploitation (as it's called in the industry) would perpetuate a deliberate and possibly fatal ignorance that I call the Myth of Away.

Let's unmask this myth with a few quick questions. Where does your tap water come from? When you're done using it, where does the waste water go? Where do your electricity and fuel come from? Where do your clothes, your transport, your medicines, the building materials in your house or apartment come from? How about each item of food in your refrigerator? Where does your garbage go? Without extensive research, we don't know where most of the things in our lives come from or go to: it's just somewhere far away.

Indigenous peoples throughout the world, especially gatherer-hunters, know exactly where their food and water come from and where their waste goes. It is no coincidence that these "natives" consume dozens of times fewer resources than we do and, in many cases, waste practically nothing. Their economies are largely regenerative: food, body and construction wastes do not travel far in time or space before the local ecology reclaims and reconstitutes them into some other useful form. Surprisingly, anthropologists have found that most such peoples (prior to devastating encounters with consumer economies) enjoyed far more free time than we do and viewed nature not as hostile, but as benign and abundant.[59]

By contrast, our industrial economy is extractive. As consumers, we buy something, use it up and throw it away. Then someone far away mines the earth to make the next item for our consumption. We don't care where it came from or where it's going, so long as we can have it when we want it. Ours is a global culture of disconnection, filled with pipes, highways and media that come from and lead to that ubiquitous, fictional non-place called "away." By denying place and linearizing the once-cyclic resource flows of natural ecology, we automatically create both depletion and pollution. These problems are built into the physical shape of our assumptions about the world.

The Space Frontier is all about "away." By exporting pollution and population and moving resources across ever greater distances, it promises only to expand the extractive economy of commercial empires, not to deepen the regenerative ecologies of life. Buying time with space resources is a siren's song. It leads directly away from the ecological knowledge we need to live harmoniously wherever we are. It beckons us deeper into exactly the troubles of growth that we would most like to avoid.

Perhaps it is ecologically functional, then, that space is so expensive to reach: if you loaded the Space Shuttle up with straw, which then magically turned into gold in orbit, you would lose money due to the transportation costs. Would it be so surprising if recycling and conservation, already economical in many cases on Earth, hint at a better path to the cosmos? We will expand on this possibility in later chapters. For now, though, it's clear that space offers no escape from our problems and probably won't even supply us economically in the short run.

Transcendence

If a few humans somehow did manage to start anew far from Earth, new social forms surely would emerge. In space, where even the Law of Gravity seems optional, other Earthly constraints may no longer bind. Who could stop a distant space habitat, for instance, from engaging in human cloning and genetic engineering? Beyond reach of today's global civilization, humanity could find endless new directions for itself: new stories, new architectures and even new bodies. Thus we come to the Frontier's third promise: transcendence.

In space, sources of water, air and soil may be scarce, but solar power abounds. Recognizing this, Tsiolkovsky suggested that space dwellers ultimately would outgrow consumption and become autotrophs, deriving their energy from the Sun as algae do. He understood, though, that any such being would need both sides of the respiration cycle if it used photosynthesis, both producing and consuming oxygen within its own body. It would function like a world in its own right: a miniature version of Earth's entire biosphere. Just as the cells of your body embody mitochondria and other evolved symbiotes,

Tsiolkovsky's autotrophs would include the metabolisms of both animals and plants.[60]

When Tsiolkovsky's hero interviews members of a species of intelligent autotrophs in his 1895 monograph *Dreams of Earth and Sky*, he learns that they originated on large planets. Upon reaching space, they gradually adapted to the weightless vacuum "just as your aquatic animals were gradually transformed into land animals, and your land animals into flying animals."[60]

Others since Tsiolkovsky have looked to technology for a much more rapid transformation. One of the earliest and most vivid post-evolutionary visions appeared in applied mathematician John Desmond Bernal's 1929 paper *The World, The Flesh, and The Devil*.[61]

In Bernal's view, science and technology were the only means to overcome the fundamental tyrannies of life. Advances in organic chemistry would provide humankind with ever stronger, safer materials, artificial foods with much more variety and nutrition than plant and animal flesh, and ultimately replace all metals and hence the need for labor-intensive mines. As processes were refined, all our physical needs would be provided for with decreasing expenditures of energy and materials (in sharp contrast to trends that have prevailed in the seven decades since his paper). Even so, he argued that human desire, "the strongest thing in the world," ultimately would prove too big for the Earth. To him, the conquest of space was inevitable.

Bernal's suggestions for how to approach this conquest have influenced the space community tremendously. His rocket boosters used beamed microwave power systems that anticipated the solar power satellite systems that O'Neill and Glaser would propose some 45 years later. His deep-space vehicles used solar sails, which we'll talk about again in later chapters. But more than any other idea, it was his space habitat that captivated the imagination of spacers over the decades. He proposed a 15-kilometer hollow sphere made of materials from the asteroids or perhaps Saturn's rings. Its transparent walls would admit sunlight, which would power its systems either using chlorophyll or photovoltaic solar cells. He likened his sphere to a phototrophic organism and, echoing Tsiolkovsky, recognized that everything within would be recycled: "the globe takes the place of the whole earth and not of any part of it, and in the earth nothing can afford to be permanently wasted." If ever space habitation had a Prime Directive, this was it.

Bernal declared that inside his sphere, "there would probably be no more need for government than in a modern hotel...Free communications and voluntary associations of interested persons will be the rule." As the human population left the planet for Bernal's stateless utopias, planet Earth, at last "free from the economic necessity of producing vast quantities of agricultural products, could be allowed to revert to a very much more natural state."

But human desire would not be content with life around just one star. Bernal foresaw not only expansion outward into the universe, but also the possibility of improving the cosmos itself for human habitation. By carefully managing the energies of each star humans visit, he wrote, "the life of the universe could probably be prolonged to many millions of millions of times what it would be without organization."

This theme resurfaced 54 years later when physicist David Criswell proposed a method to increase the lifetime of the Sun. He called it "stellar husbandry" or "star lifting."[54] The technique involved siphoning mass from the Sun and using it to build new worlds. Enough new land would be produced to house some 10^{17} human beings. For every person now alive, Criswell's civilization would have over 12 million people living at a level of affluence undreamed of today. Meanwhile the Sun, stripped of most of its mass, would shrink down to a white dwarf. By surrounding this shriveled star completely with habitats, Criswell's civilization could provide each citizen with some 200 kilowatts of power and light—comparable to the power level I will compute for gaiomes on page 151. White dwarfs are so hot and dense that they take tens of trillions of years to cool off. Criswell proposed to extend even this long lifetime another factor of ten by forming new white dwarfs out of reserved hydrogen, or pouring it in a steady trickle onto the tiny Sun's surface, where it would instantly fuse to provide light and heat. Thus his scheme would extend the Sun's tenure as a power source for life by a factor of ten thousand.

Even such grandiose conquest was not enough for Bernal. "Normal man," he wrote, "is an evolutionary dead end."

Bernal imagined improving on the frailties of the human form through what can only be described as the ultimate in elective surgery: removing the brain entirely from the body and placing it in an indestructible cylinder. Inside, machinery would circulate all the nutrients necessary to keep the brain alive, and the nerves would be grafted to conduits leading to external sensors and appendages. We would become immortal cyborgs, more machine than human, capable of living in open vacuum and seeing directly all the spectra of light now invisible to our eyes. We would communicate by radio-telepathy, eventually evolving such close associations with each other's thoughts as to become hive minds. "The new life would be more plastic," wrote Bernal, "more directly controllable and at the same time more variable and more permanent than that produced by the triumphant opportunism of nature."

These writings evoke a sense of rage at the happenstance nature of existence. Rather than bow to random mutations and evolutionary selection pressures, Bernal would have us take our destiny into our own hands, shaping ourselves and all the environments we encounter according our own conscious desires. Clearly, space would give us room to do that.

But what has shaped our desires in the first place? In all cases, it was our environment, whether natural or built. We have evolved together with millions of other species, each turning point in our evolution shaped by the presence of other beings that were themselves shaped by our emergence. We belong to a natural democracy that long succeeded in matching our desires to our environment. We like greenery, nice weather and companionship not only because these provide for us, but also because we were, for most of our history, integral to the systems that maintain them. We once were native to Earth.

So it was for the first 20 pages of our story as a species. But on the 21st page (the millionth in the Book of Earth), the dawn of writing touched off a process that gradually distanced us from immediate sensory experience.[62] As these have evolved into ever more immersive media such as film and video, we find we can manufacture experience, shaping desire and broadcasting it for profit. Not surprisingly, our desires increasingly come to be dominated by manufactured items, themselves extracted from the earth and discarded when spent.

As Bernal foretold, our chemistry has indeed advanced to the point where we manufacture tens of thousands of chemical compounds and alloys that are seldom if ever found in nature. Not surprisingly, ecosystems have not yet evolved ways to reclaim many of these substances. A discarded plastic spoon or carpet may sit in a landfill for many thousands of years, during which time its wooden counterpart could have cycled its vital nutrients through the ecology (including plants, animals and people) hundreds of times.

Building artificial bodies for ourselves in space would set us on the most lonely, arduous path imaginable. Not only would we have to re-invent a solar economy that took nature billions of years to develop, we would literally have to chop off the larger parts of ourselves and abandon them forever. I'm not merely talking about our bodies. Even a cursory study of living nature shows that our identity does not end at our skin. Life evolves simultaneously on all scales, from the sub-cellular to the largest ecosystems. Thus our being encompasses the taste of ripe berries, the warming rays of dawn, the brush of a breeze, the blue sky, the quenching downpour, the scent of wildflowers, the voices of crickets and birds, the lay of the land and the regenerative services of the soil microbes. All these parts of us preceded our hands and intellects by millions of years into every Earthly frontier.

The regenerative economy, the community of life, the truest friends we ever had: none of these await us in space. What's left of us once we transcend them?

The Finish Line

When Neil Armstrong's small steps on the Moon ultimately won the race for NASA, the Final Frontier seemed wide open at last. How strange, then, that over the next few years, the nation turned away from space. It wasn't simply that Presidents Nixon, Ford and Carter lacked Kennedy's vision. After all, the bold space proposals of Ronald Reagan and George Bush Sr. have also come to naught. Nor was it simply that space somehow was packaged in a boring manner. The problem was that to most Americans, the space frontier never opened at all.

Turner recognized that the central element of the American frontier was land that anyone could reach and claim with a plow and a gun. Its accessibility made the frontier something that nothing could repress: not legislation nor civic boundary nor even instances of compassion for indigenous people. Because it was relatively easy to reach, it physically absorbed and sustained many millions of settlers over the course of several centuries.

Space is not like that. You cannot just walk there. The Apollo missions, at a cost of billions of dollars, managed to put only twelve men on the Moon for a total of twelve and a half days. Only the best pilots and the most promising scientists in the best possible health could become astronauts—but only if they also looked good on TV. Under these circumstances, most people had a much better chance of winning a major lottery than ever going to space, and the lottery cost only a buck to play. The dry, technical, stressed-out life of the astronaut promised nothing to the average citizen.

If space is ever to achieve any tangible meaning for the rest of us, we must find a way to make it much easier and less perilous to reach. I discuss how this might be accomplished in the next chapter. Even then, though, space won't be a frontier and we would be foolish to try to make it one.

The world's heroic government space programs, built on the myth of "away," have fallen far short of our hopes for expansion, escape and transcendence, and always will. The hopes themselves are wrong. If we try to propagate our pattern of obligate growth throughout the cosmos, we would succeed, at most, in greatly expanding the anxiety and misery it already exports. If our extractive economy wrecks Earth for us via resource wars or environmental crises, it also leaves us even less prepared to survive anywhere else. If we surgically sever all relations with our world and our bodies, we'll find ourselves friendless, estranged and hollow.

Yet space is rich in promise: not as a venue for conquest or escape, but as a challenge to evolve, a challenge not for one isolated species but for whole biomes. Life should try living beyond Earth, just as it tried to grow on land many times over until at last finding its stride through massive coevolution in the Cambrian period and beyond. Gaiomes are worth building in their own right, if only just to see what happens next.

We have tried building sterile, submarine-like habitats such as Mir and the International Space Station (ISS). While efficient for short trips with a few people, in the long run that's the hard way. It will always be dangerous, expensive, depleting and elitist. It ignores the great challenge of evolution because it excludes the vast community of people and species with whom we've evolved.

Alternatively, we could bring as much Earth life as we can along with us into space and do all we can to help it thrive. But first we have a lot to do and a lot to learn. We need to let life become a circle again here on Earth. We need to do away with "away." We need to re-learn the art of locality, of being native to a place and aware of its cycles. Nowhere will we need these vital skills more than in the deep vacuum of space, where, as Bernal said, "nothing can afford to be permanently wasted." Only when our relationship with Earth and all its species is mutually secure, abundant and joyous will we find ourselves—our larger, ecological selves—able to experiment meaningfully with living beyond.

Chapter 2

Space for Everyone

"Perhaps evolution and progress *can* converge in a single process as we reenter the free community of sovereign species as equal (not dominant) members, and devote our attention to the welfare of that entire community. Perhaps we may eventually even realize our Promethean dream of being space explorers...though not as Star-Trek colonists and cosmocrats, but as children waking from sleep, ready to go outside and play."
—Richard Heinberg[63]

Eventually we will know enough about living sustainably on Earth to try living in space. When that day comes, people will attempt it, just as our ancestors back to the first prokaryota tried to live beyond their immediate horizons.

In the meantime, we should continue to explore the solar system using such modest instruments as telescopes and robot probes. Judging by their relatively small budgets, these investigations were a mere afterthought in the von Braun plan. But they remain the most enduring scientific achievements of NASA and the world's other space programs. They have mapped many new worlds over the past few decades, making ours by far the greatest age of exploration in history. Yet untold millions of minor planets remain uncharted.

Space exploration has made us aware of threats such as ozone depletion. Had a few more decades passed before satellite data revealed that our refrigerants and aerosol propellants were destroying the ozone layer, we would now be in much deeper trouble than we presently are. Large, Earth-approaching asteroids pose a threat, too, and our surveys of them will be nearly complete within a decade. Increasing awareness of our cosmic situation can't help but improve our odds of long-term survival.

As we become native both to Earth's ecologies and the wider cosmos, those of us who are serious about space travel should strive to make it available to everyone. Doing so would start to bring life's strengths—multiplicity and diversity—to bear on the challenges of cosmic evolution.

Unfortunately, space is hard to reach—just ask anyone at NASA. From the Agency's earliest days, space travel was thousands of times more dangerous and hundreds of thousands of times more expensive than commercial air travel (we'll get to the actual numbers in a few pages). Space began as the domain of giant national budgets, technological elites and brave heroes: the stuff of empire. But even at the beginning, NASA's employees and fans looked forward to a time when space would open up for the rest of us, just as the sky gradually did for millions of passengers a mere fifty years after Kittyhawk.

Fifty years have passed since Sputnik ushered in the space age and NASA has since spent over $600 billion.[64] Yet fewer than five hundred people have been to space. Unlike air travel, which became five thousand times safer in the USA from 1926 to 1976,[65] space travel has not become any safer or cheaper. Why is that?

Commercial space travel must face and solve three areas of challenge before it can happen. These include physical, financial and environmental concerns. The fact that space industry procedures and thinking have changed little in all three areas for nearly fifty years suggests that solutions will not much resemble today's industry at all. Before we get into the specifics of how passenger space travel might actually work, let's briefly examine each challenge.

The Physical Challenge

It is astonishingly difficult to reach space and return safely. First, your spacecraft must break free of the atmosphere. Then it must build up many times more speed to remain in orbit. To return home, it must re-enter Earth's atmosphere, enduring tremendous heating and crushing decelerations. Finally, it must slow to a safe speed and land with precision. Each step is daunting. To date, only rockets have managed even the first step.

Rocketing Into Space

In theory, a rocket is simple: propellants combine and burn in a combustion chamber. The resulting exhaust finds a convenient exit through a bell-shaped nozzle. The exhaust goes one way; the rocket, obeying Newton's third law, picks up speed in the opposite direction. The hotter the combustion, the faster the exhaust, and the faster the rocket will go for each tonne of propellant burned. Liquid fuels (such as kerosene) burn much hotter than solids, so the most efficient rockets are liquid-fueled.

In practice, modern liquid rockets have proven difficult to perfect. They consume enormous amounts of propellant in seconds, so pumps must work at very high pressures. In the combustion chamber, injectors must bring the propellants together efficiently over a wide range of pressures. The intense heat of combustion tends to melt the chamber and nozzle, so cold propellants are cycled around them for cooling. The pressure, heat and vibration tend to warp and crack the materials involved. It took von Braun and teams of hundreds of scientists nearly 30 years and many thousands of tests to solve these problems. Most rockets trace their lineage back to similarly stringent efforts. It's not called rocket science for nothing!

A passenger rocket might take you straight upward with an acceleration of three gees, about as much as you would feel for a second or two on a looping roller coaster. But to clear the atmosphere, your rocket must fire for a solid minute. If that sounds stressful, it is, but most healthy people could handle it with minimal preparation. By the time your rocket cuts off, you would be 45 kilometers above the ground, flying upward at a kilometer per second.

At this altitude, there's so little air left above you that the sky has drained away to black. Stars glare, unblinking, above a thin, blue horizon.

So there you are at four times the altitude of a commercial jet, soaring spaceward at Mach 3. With no rocket power, you and your vehicle are coasting together: you're weightless. All the while, Earth's gravity is slowing your ascent until, ninety seconds after engine shut-off, you and your capsule come to a stand-still at 100 kilometers altitude. If nothing else happens, in another two minutes gravity will bring you crashing back to Earth.

Orbits and ΔV Budgets

What if you could somehow dodge the Earth before that happened? Suppose you got moving parallel to the ground, so that its round surface started to curve away from you. With enough speed (about 8 km/s, or Mach 25), you would fall freely without losing altitude. Your trajectory would make a circle about the globe: welcome to orbit!

How much propellant would you need to get there? In space, distance alone won't tell you much: establishing orbit at 100 km requires nine times more speed than merely reaching that altitude. Speed, then, provides the best measure of performance, especially because it can be related directly to the speed of the rocket's exhaust gasses. The total of all of the changes in speed that a rocket makes to reach its destination is called the ΔV ("delta vee").

There is a dazzling array of orbits to choose from. Low Earth Orbit (LEO) extends from 100 km, where the atmosphere is just thin enough to allow a full orbit, up to 1,000 km. If your spacecraft starts out faster than 8 km/s, the far end of its orbit will swing out away from Earth, losing speed against gravity until it reaches its apogee (highest point) on the opposite side of the world. Its orbital path traces out an ellipse with Earth at one focus. If you give your spacecraft enough ΔV, 11.2 kilometers per second, its path will not close like an ellipse, but instead will trace out a trajectory of no return: a parabola if your ΔV exactly matches the escape speed, or a hyperbola if your ΔV exceeds it.

Ground controllers often "park" a spacecraft in a low circular orbit, then later on give it some additional speed to put it into an elliptical orbit. Finally, at apogee, they'll add a little more speed to circularize the orbit. This is how they launch communication satellites into 24-hour geosynchronous (GEO) orbits. In GEO, satellites such as DirecTV remain above the horizon day and night in their target markets. All of the changes in speed needed to send it there add to its ΔV budget.

Propellants and Exhaust Speed

Cars and airplanes use the oxygen freely available in air to burn their fuels. In the near-void of space, oxygen is much too sparse to burn anything, so spacecraft must carry all of their propellants; these usually include both a fuel and an oxidizer.

A car's fuel economy tells you how much gas or diesel it will need to drive

Maneuver	ΔV, km/s	Propellant fraction	
		Kerosene	Hydrogen
Earth surface to escape	11.7	98%	95%
Earth surface to LEO	9.2	95%	90%
LEO to Moon's surface	6.3	88%	79%
LEO to nearby asteroid	4.0	74%	63%
LEO to Mars	3.7	71%	60%
LEO to GEO	3.5	69%	58%
LEO to escape	3.2	66%	55%

Table 2.1: Maneuvers, their ΔV requirements and the propellant mass fraction required to complete them using a LOX oxidizer and either kerosene ($v_e = 3$ km/s) or hydrogen ($v_e = 4$ km/s) fuel.

a certain distance. Similarly, a rocket's exhaust speed will tell you how much propellant it must carry to reach a certain ΔV.

The first two columns of table 2.1 list several trips and the combined ΔV of all maneuvers required to complete them. The first row shows the escape speed that we talked about in the last section, plus 500 m/s lost to atmospheric drag during ascent, for a total of 11.7 km/s. The next row shows the ΔV required to reach LEO from the ground.[66] The remaining rows show the ΔV involved in transferring from one orbit to the next. Note that if you add the 9.2 km/s to reach LEO from the ground to the 3.5 km/s required to go from LEO to GEO, you get 12.7 km/s—more than the escape speed from the ground! Getting into and out of that parking orbit in LEO costs some extra ΔV. As happens on Earth, extra stops can add to a trip's cost.

To reach a ΔV equal to the exhaust speed, 63% of your rocket's mass at launch must be propellant. To reach twice the exhaust speed, 63% of what remains after burning that first 63% must also be propellant, for a total *propellant mass fraction* of 86%. Clearly, the faster a rocket's exhaust, the less propellant it will need to reach a certain speed.[67]

The exhaust speed ultimately depends on the choice of propellants. Kerosene burning in liquid oxygen (LOX) has an exhaust speed no higher than 3.5 km/s. The most efficient propellant combination in use is liquid hydrogen and LOX, which combine to form water with an exhaust speed of up to 4.5 km/s.

The last two columns in table 2.1 show how much propellant your rocket will need to reach each ΔV. For these calculations, I've used real-world exhaust speeds (3 km/s for kerosene and 4 km/s for hydrogen) that account for lower performance in the atmosphere and other losses involved in making the engines reusable. Let's skip the first row for now. Reading across the second row, observe that it takes 9.2 km/s to reach LEO from Earth's surface.[66] Using kerosene and LOX (third column), 95% of your rocket's mass must be propellant. If you use hydrogen and LOX (fourth column), only 90% of your rocket's mass need be propellant.

How much cargo can your rocket carry? That depends on how much is left over after subtracting the propellant fraction (which we know from the table) and the *dry mass fraction*. Dry mass includes the tanks, engines, guidance systems, landing gear and the structure that holds them all in place—but not the propellant, passengers or cargo. Modern liquid rockets tend to have dry mass fractions in the 5–10% range.

Suppose that our LEO-bound rocket in table 2.1 has a dry mass fraction of 6%. If it burns kerosene, it's out of luck: adding the dry and propellant fractions (6% + 95%) already puts it over its mass budget of 100%. To reach orbit, it needs a hotter-burning fuel. Hydrogen would suffice: burned in LOX, it would put a cargo of 100% − 90% − 6% = 4% of the total mass into orbit.

Let's have a look at that first row, which describes the parabolic escape orbit briefly mentioned in the previous section. With a dry mass fraction of 6%, the propellant fraction for both kerosene and hydrogen-fueled rockets goes over budget. It is very hard to build anything with a dry mass less than 6%. Yet hundreds of scientific probes have managed to escape Earth's gravity entirely. How? They used multiple stages.

Staging

Imagine that you could only fuel your car once, at the time of purchase. For a decade's driving at reasonable fuel economy, it might require something like 25 tonnes of fuel: 10–20 times the mass of your car. Clearly, lugging around all that extra fuel would take a serious toll on your car's economy for the first few years. This is just the situation that rockets face: they can't refuel on their way to orbit.

Still, they can lighten their load by dropping their massive propellant tanks one after another as soon as each is empty. Designers have used this method to coax better performance out of rockets since the 16th century. From high-flying fireworks to satellite launchers, most rockets throw away spent stages as they ascend. This frees the spacecraft from the burden of having to lift the "dead weight" of massive, empty tanks all the way to orbit.

For example, two kerosene-burning stages could reach LEO with a propellant fraction of 78% in each stage. If the dry mass again accounts for 6% of each stage, then 16% of each stage is cargo. The second stage is the first stage's cargo. That puts 16% of 16%, or 2.4% of the launch mass into orbit: a great improvement over the single-stage launcher in the previous section, which could not even reach orbit with kerosene fuel.

If two stages are good for the mass budget, more stages are better. Deep-space probes typically use three or four stages to reach the high ΔV values needed to reach other planets.

Re-entry

De-orbiting a spacecraft from LEO is easy: just fire rockets in the direction of travel to drop the perigee (lowest orbital point) into the atmosphere; drag does the rest. However, surviving the journey is another matter. A returning space capsule re-enters the atmosphere at Mach 25, fast enough to form a white-hot shockwave in front of it. This can heat the surface of the spacecraft to temperatures as high as 1,500 K—well past the melting or ignition point of most materials.[68]

There are some tricks to surviving re-entry. Spacecraft can enter at a shallow angle, losing speed gradually in the extremely tenuous upper reaches of the atmosphere. Even so, they must be built strong enough to take the stress of deceleration, typically three to ten gees. Surfaces exposed to heating need to be wide, thick and heat-resistant. Thermal protection can be as complex as the Space Shuttle's mixture of carbon-carbon reinforced leading edges and air-puffed ceramic tiles, or as simple as the cork shields of some Chinese space capsules. Cork is one of many materials that char and ablate (flake apart) during re-entry, carrying heat away like a meteor trail. Although they are cheap, ablative re-entry shields cannot be re-used like the Shuttle's tiles.

Airplane-like spacecraft such as the Shuttle must re-enter belly first to spread the heat over as large a surface as possible. On the Shuttle, this orientation requires constant computer control to maintain. Burt Rutan's *SpaceShipOne* design (which we'll discuss in detail soon) uses a hinged wing to hold the vehicle at the correct angle without computer control during re-entry. However, this design has only been tested at about Mach 3, or about 60 times less total heating than it would face coming back from orbit.

Re-entry has been a major obstacle to routine spaceflight for decades. Accidents during this phase are common, having killed the crews of the Soyuz 11 in 1971 and the Space Shuttle Columbia in 2003. Precision steering during this critical phase also poses problems. During Soyuz TMA-1's descent in 2003, for example, the guidance system failed, dropping the capsule some 450 km short of its planned landing site.[69]

Landing

Once a spacecraft has made it through the upper atmosphere, it must navigate the lower atmosphere to a safe landing. By far the simplest method is to use parachutes. Early American space capsules used this technique, splash-landing in the ocean to be recovered by ship and helicopter. To this day, Russian Soyuz capsules still use parachutes and recovery crews, landing with rocket assist on land rather than water. Parachutes get the job done within a tight mass budget, but they offer neither precision nor especially high reliability. For example, a parachute failure on Soyuz 1 (in 1967) destroyed the spacecraft, killing its pilot-cosmonaut.[70]

The Space Shuttle glides to a landing on a runway, which gains it some degree of precision but adds the complexity and weight of landing gear and wings. The Shuttle also requires a very long airfield to land.

Other landing schemes such as tail-first, rocket-only descent and hovering with autogyro-style helicopter blades have been proposed, but never fully tested. These concepts have the advantage of not requiring large runways to land, making spaceports less costly. They also offer much higher precision than parachutes.

Safety

Given the large number of physical challenges that rockets face, it's not surprising that they are dangerous. Table 2.2 compares their safety record to that of commercial aircraft. The final column divides the number of flights (second column) by the number of crashes (third column) to give your odds of surviving a flight. For example, when you board a commercial jet, your odds of surviving the trip are overwhelmingly favorable: 547,000 to 1. You could

Vehicle	Activities	Lost	Ratio
Aircraft			
Commercial Jets[71]	430,000,000 flights	786	547,000:1
Private planes[72]	27,800,00 hours	345	82,600:1
Spacecraft			
Space Shuttle	118 launches	2	59:1
Soyuz	97 launches	2	49:1
V-2 Rocket[23]	1,027 launches	79	13:1

Table 2.2: Safety of aircraft vs. spacecraft as a ratio of the number of flights, flying hours or launches to the number of vehicles lost in fatal accidents.

fly on commercial jets every week of your life without significant risk. If you board a Space Shuttle, however, your odds drop to 59:1. A year of weekly Shuttle flights would probably kill you.

During the first two decades of aviation, when aircraft were nearly as dangerous as spacecraft are today, the public rightly feared to fly. This tended to limit demand for new aircraft. American manufacturers could not afford to break away from the pack and build safer planes: that would price them out of the market. So together they asked government to step in and regulate the industry. The Air Commerce Act of 1926 set up agencies to license pilots, certify aircraft, enforce safety rules and investigate accidents. Over the years, as designs, maintenance and navigation improved, commercial aviation became 2,000 times safer, and the number of air travelers grew by a factor of 30,000. Eventually the regulatory agencies merged to become the Federal Aviation Administration (FAA).

When a new airliner rolls out for the first time, it must perform some 1,000 test flights to earn its passenger certification.[73,74] It also must conform to numerous guidelines for construction that have been found over the past century to contribute to fault tolerance, or the ability to survive failures that sometimes occur during normal operation. Commercial aircraft, for example, are built to survive an engine failure, and test-flown multiple times with an engine shut down to verify this. Stringent regulation has indeed made air travel safe and cheap, but it also has made airplane development extremely expensive.

By aviation standards, neither the Shuttle nor the Soyuz would qualify as operational. With 118 and 97 launches apiece, (Table 2.2) they are merely experimental vehicles barely a tenth of the way through their test programs. Both lines of spacecraft have suffered numerous unanticipated failures, including two fatal crashes each. Had these been real test programs, neither vehicle would have earned a passing grade.

I remember watching the news with incredulity on July 4, 2006, as flight controllers gave Shuttle mission STS-121 the go-ahead to launch. This decision directly overruled the judgment of safety officers concerned with new cracks in the insulating foam of the external fuel tank. I suppose I could try to defend the foam flecks as innocuous—had they not played a major role in destroying STS-107 two flights earlier. The fact that NASA had already spent $1.3 billion trying to fix the problem (a significant fraction of the cost of a new Space Shuttle) made the situation especially hard to accept.

The space industry has barely begun to face, let alone understand the factors that contribute to passenger safety. The challenges are physical, not just managerial, and their solutions will require much more testing than we've seen to date. NASA has spent $145 billion on the Shuttle thus far and NASA chief Michael Griffin has admitted that it "was not the right path."[75,76] The Shuttle is a bird of compromise: built to lift large satellites, keep astronauts alive for weeks on end, involve contractors from key congressional districts and perform complex, military maneuvers per special arrangements with the Air Force. Future efforts will need at least ten times more testing and, to avoid breaking the bank, much narrower focus.

Even so, at today's technology level, one could anticipate losing several vehicles early in a 1,000-launch test program. Losing a pilot automatically threatens a program with delay, shut-down and lawsuits. But why have pilots? Rockets typically move at speeds where conditions change much faster than human reaction times can track, so their flight systems typically involve extensive computer control. A vehicle without a pilot in the loop could build up an impressive number of flight hours, and hence a much larger assurance of safety, without ever putting a person at risk. Blue Origin, Amazon.com Founder Jeff Bezos' secretive rocket start-up, seems to be following just this strategy. Over private land, its vehicle takes off and lands vertically entirely under computer control.[77]

Testing, fault-tolerance, focus and automation are the watch-words for passenger safety in space. But what will a ticket cost?

Vehicle	Cost ($M)	Staff	Flights	Passengers	Ticket ($)
Shuttle	1,500	10,000	1/yr	6	200M
747	250	50	300/yr	368	400

Table 2.3: Cost and performance variables for the Space Shuttle and the 747-400 passenger jet.

The Financial Challenge

It takes an enormous amount of propellant to reach orbit: anywhere from 20 to 100 kilograms per kilogram of cargo, depending on staging and other factors. Yet the price of propellants isn't so bad: about $0.50–$1.00 per kilogram. Airlines are profitable at ticket prices around five times fuel costs, so visionary rocketeers have long looked forward to the day when sending cargo to space might cost only $50–$500 per kilogram delivered to orbit.

Unfortunately, space travel does not enjoy anything like airline efficiency. Over the past 25 years, there have been 118 Shuttle launches, at a total cost of $145 billion.[75] Each launch, then, cost $1.2 billion dollars. If the six passengers aboard each flight had to pay their own way, their tickets would have come to $200 million apiece. Since a Shuttle can carry about 25 tonnes of cargo, it costs about $50,000 to lift a kilogram into space, of which propellants account for only about $71. Clearly, propellant costs have almost nothing to do with the Shuttle's expense.

What would it take to knock a few zeros off of the Shuttle's ticket price? Table 2.3 provides some guidance. For starters, you can save a factor of 200 by trimming the support staff from 10,000 down to the 50 or so needed to service and fly the Boeing 747 airliner.[78,79] What about the frequency of flights? The fleet of five shuttles has averaged less than one flight per vehicle per year. No airline could survive if it flew each plane just once a year. Let's bump that flight rate up to 300 per year. Since the Shuttle can lift 25 tonnes, and each passenger with life support should weigh no more than a tonne, its passenger capacity could be around 25 instead of the 6 (plus a pilot) that it carries now. Of course a Shuttle costs more to build than a commercial jet: $1.7 billion vs. $250 million for a 747.[80] Still, airline-like efficiency could divide the non-fuel costs of operating a Shuttle (about $45,000/kg of cargo) by a factor of $200 \times 300 \times (25/6) \times (250/1,700) = 36,754$. That comes to $1.36 per kg lifted to orbit. Propellant costs then become significant at $71 per kg of cargo or passengers (plus life support) delivered to orbit. Round up the total to $73/kg, multiply by five to account for airline-like operating expenses and profits, and you end up paying $365/kg to orbit, or $365,000 for a round-trip as a passenger.

To summarize: travel aboard the Space Shuttle is expensive not because of the huge amount of propellant it consumes, but mainly because it doesn't fly often enough and it takes too many people to maintain.

The Market for Passenger Space Travel

If a commercial passenger rocket could be designed for simple maintenance and frequent launches, it could bring a ticket to space down to about the cost of a swank high-rise apartment. But how many of us have that sort of money lying around? Amazingly enough, a large, thriving aviation market already exists at nearly this ticket price.

The wealthiest business travelers can get a sixteenth share in a private jet for around a million dollars a year. That's about 50 flight hours a year, with neither the crowds, hassles and delays of airline travel, nor the inconvenience of scheduling, staffing and maintaining a jet. To fly from New York to Los Angeles, the total cost works out to $100,000 for a 5-hour flight. It seems likely that customers would pay even more to trim flight times down to the 40 minutes that ballistic rockets could offer. This market is not small. With up to 350 flights per day, NetJets, the leader in the fractional ownership business jet category, was effectively the eighth-largest airline in America in 2001.[81]

Then there are the thrill-seekers and the operators who aim to satisfy their desire for the space experience. The expense, difficulties and dangers they face are unparalleled anywhere else in the travel industry, and the term "space tourist" doesn't do them justice. I prefer to call them space adventure travelers.

Dennis Tito, the world's most famous space adventure traveler, began his career at NASA in the 1960s. It wasn't long, however, before he found more lucrative employment applying his math skills on Wall Street. Three decades later, fortune in hand, he approached Space Adventures Ltd. of Arlington, Virginia to be his travel agent for a flight into space.[82] NASA turned down their request, so Space Adventures got him a seat aboard a Russian rocket to orbit. After training for months with cosmonauts in Star City, Moscow, Mr. Tito rode a Soyuz capsule to the International Space Station (ISS) in April, 2001. For his training and ten-day stay aboard the ISS, he paid $20 million. As of mid-2007, four others have followed in Mr. Tito's footsteps.

NASA managers were appalled with Mr. Tito's flight on at least two levels: first, because they had no say in the matter, and second, because the "tourists" got far more publicity than all of the research conducted aboard the ISS. "It's like we're building the pyramids of Egypt in orbit," Space Station Program Manager Tommy Holloway told the press, "and nobody's watching."[83]

NASA's Space Pharaohs are kidding themselves. While it's true that the ISS and the Great Pyramid of Giza both occupied thousands of workers for decades,[84] great monuments are not built on massive budgets alone. The Pyramids have stood for over 4,500 years; millions of people have visited them in safety. There's no safe way to visit the ISS and, without constant resupply, Earth's tenuous outer atmosphere will pluck it out of orbit in less than a decade. The space station is a lousy monument, but for now it's the only passenger destination in space. That will change.

Tourism (including all long-distance travel) is very big business. At $3.3 Trillion USD or 11% of the Global Gross Domestic Product, it's a contender for the largest industry in the world.[85] Moreover, it's an industry that is used to enormous capital investments. Market surveys suggest that up to 10% of the general population would spend a year's salary to travel in space, so there is already a large, untapped market for passenger space vehicles and orbital destinations.[86]

The X-Prize ...

Despite the tempting economic incentives, technology remains a hurdle for those who would invest in space transportation. But investment is a prerequisite for technology development. In the early years of aviation, cash prizes for technical achievement helped to break this stalemate. Perhaps the most famous among these was the $25,000 Orteig prize that motivated Charles Lindbergh and his sponsors to make the first transatlantic flight in 1927. This achievement galvanized public interest in flying. According to the U.S. Centennial of Flight commission, "only 5,800 passengers had flown in 1926; this leaped to 417,400 four years later."[87]

In 1994, physician and serial space entrepreneur Peter Diamandis was reading *Spirit of Saint Louis*, Lindbergh's Pulitzer-winning account of the flight, when it dawned on him that space travel could use its own prize. This was the birth of the X-Prize Foundation, which in 1996 announced a $10 million prize for the first private team that could launch the same 3-seat spacecraft to 100 km altitude twice in two weeks. It was later renamed the Ansari X-Prize in honor of its largest contributor, Anousheh Ansari, who would became Earth's fourth space adventure traveler (see page 179).[88]

Over the following decade, 23 teams emerged from seven countries to compete for the prize. On October 4, 2004, Mojave Space Ventures won it. By this time, not many were surprised, as this team was led by famed airplane designer Burt Rutan and funded to the tune of $30 million by former Microsoft co-founder Paul Allen. *SpaceShipOne*, Rutan's winning design, was carried aloft by an airplane (called *White Knight*), and released at high altitude. Rockets burning solid rubber in liquid oxygen then roared to life and carried its lone pilot (plus ballast in the other two seats) up to the winning altitude.

...and Beyond

As Diamandis had hoped, the success of the X-Prize competition immediately opened the hearts and purse-strings of private finance.

A week before *SpaceShipOne's* winning flight, Virgin Airways founder Richard Branson announced plans for Virgin Galactic, a spaceline of commercial 8-seaters that would repeat its X-Prize winning ride with paid passengers.

In a nearly simultaneous announcement, Las Vegas hotel tycoon Robert Bigelow unveiled plans to award a $50 million purse, called *America's Space Prize*, to the first 5-seat vehicle to transport passengers to and from Bigelow Aerospace's inflatable space habitat modules. The rules stipulate that the vehicle must be privately funded (i.e. not built on government contract), 80% re-usable, able to reach 400 km altitude, and capable of docking with Nautilus. And it must make two deliveries within 60 days by January 10, 2010.

All of this is very exciting for prospective space adventurers. Mainstream space tourists, however, will have to wait until the technology matures before they can buy tickets for themselves and their families.

Are Today's Spacecraft too Cheap?

Many in the space industry have concerned themselves with bringing down the price of spacecraft. However, as we saw on page 56, the ticket price has a lot more to do with flight rate and staff size than the cost of the vehicle. Today's rockets are, if anything, too cheap—and it shows. They still have far too much heritage from the V-2 and other expendable missiles. In table 2.2 (page 54), for example, you can see that the Shuttle is barely five times safer than the Nazi war rockets, which were subject to sabotage, air raids and supply problems.[23] Who in their right mind, no matter how rich, would take their family on a modern rocket?

Of course, it's easy to spend any amount of money without getting results. The question is: what's the minimum amount required to succeed in a focused program? To get a very rough estimate, let's compare present rocket development costs per tonne to those of aircraft. We'll leave out the mass of propellants, which are negligibly cheap compared with the labor required to design, tool, fabricate, test and certify vehicle hardware. We should also ignore the mass of the cargo and passengers, because in aviation, they cover costs only over the course of thousands of flights. What's left is dry mass.

Table 2.4 shows the dry mass of three commercial aircraft and two spacecraft, together with their development costs and resultant cost per tonne. For example, the 180-tonne Boeing 747 first flew in 1969 and cost $11 billion (2005 dollars) to develop.[89] Dividing this cost by the dry mass gives a value of $64 million per tonne. The Airbus A380, which first flew in 2005, cost slightly less per tonne to develop.[90] Figures are sketchier for the Eclipse 500, a 5-seat jet certified for commercial use in 2006. A company press release states that Eclipse Aviation received a total of $320 million in venture funding to bring the jet into being.[91] That comes to about $200 million per tonne, nearly four times as much as the Airbus.

You might think that the Space Shuttle, with its much tougher jobs of achieving orbit and surviving re-entry, would have cost more per tonne to develop than any jet. Not so. The 220-tonne spacecraft became operational after an expenditure of $33 billion—or $150 million per tonne.[92] That places it well within the range of commercial airplane development.

Compare these costs with Paul Allen's X-Prize winning vehicle *Space-ShipOne*. Together with its carrier aircraft White Knight, its per-tonne cost was less than a fifth of what Boeing spent developing the 747. As you might expect, *SpaceShipOne's* inaugural rides were bumpy and fraught with peril. On his first two flights, Pilot Mike Melvill experienced uncontrolled rolls, and on separate descents, both he and fellow test pilot Brian Binnie encountered "alarming" buffeting. It would be a serious mistake to assume *SpaceShipOne* is any safer than the Space Shuttle, which killed 14 people in its first 113 launches.

Vehicle Name	Mass (tonne)	Development Cost	
		Total ($M)	$M/tonne
Aircraft			
Boeing 747	180	11,000	64
Airbus A380	280	15,000	53
Eclipse 500 Jet	1.6	320	200
Spacecraft			
Space Shuttle	220	33,000	150
SpaceShipOne	3	30	10

Table 2.4: Comparison of vehicle development costs per tonne.

Billionaire Richard Branson has announced plans to build a commercial 8-seat version of Allen's *SpaceShipOne*. In all likelihood, this will cost him at least as much per tonne as Boeing spent on the 747, and perhaps even more than Eclipse is spending now. With twice the weight and 16 times the development cost per tonne as *SpaceShipOne*, he could easily end up spending a billion dollars on design, simulation, development, testing and certification before he has a safe, reliable vehicle.

The Beverage Can Rocket

If spacecraft are built so cheaply, then why does a ticket to orbit still cost $20 million? Depending on the type of vehicle, there are two economic reasons. First, as we saw in the case of the Space Shuttle, the world's only re-usable orbital launch system, the flight rate is much too low and the staffing far too high to support a low ticket price.

Second, all other rockets (for example, the Soyuz) are expendable, meaning they're only used once. Nearly everything—the fuel tanks, piping, turbomachinery, engines, hull, thrusters and fairing (nose cone)—gets thrown away with each launch. Even the Space Shuttle is not fully re-usable: it requires a new Main Fuel Tank and pair of Solid Rocket Boosters for each flight, as well as replacement thermal tiles.

Rocket parts aren't cheap. To save weight and survive the stress and heating of atmospheric entry, they require intricate machining and exotic materials such as titanium. Why throw something so expensive away after just one use?

Today's orbital boosters trace their heritage not to aircraft, but to the ultimate one-use product: war missiles. Each one is expensive and expendable, but as we saw in the last section, their development programs have remained surprisingly cheap relative to the problems they must solve. To date, no one has taken the time or spent the money necessary to give space vehicles aircraft-like fault-tolerance and durability.

Expendables still appeal to the space industry because the price of throwaway items often can be offset substantially with large production runs. Such *economies of scale* are the secret to the affordable price of all manufactured goods, from beverage cans to laptop computers. Mass production has been so successful that dry goods have become disposably cheap in our society, so much so that it contributes significantly to pollution and resource exhaustion.

Economies of scale bring down price by taking the cost of solving a problem and distributing it among many buyers. If you want a sealed beverage can, for example, you could hire a machine shop to build it. By the time the designer and machinist are done with it, the can could easily cost a thousand dollars. While the shop has the tools set up for the job, it could make ten more cans in a small fraction of the time it took to make the first. The price per can then would drop to maybe a couple hundred dollars each (the so-called *learning curve* effect).[93] For larger orders, the shop might streamline its process or make a custom tool, resulting in further cost reductions. But what happens if you want a million or even a billion cans? At this point, you don't need a machine shop, you need a factory, and you will probably have to pay for large, custom tools. These might cost millions. But your high capital cost gets spread over a billion cans, dropping the cost per can to mere pennies. A factory can also buy raw materials in bulk at a discount, dropping costs even further.

Think of it: two nearly identical cans. One costs a thousand dollars, the other costs a few cents.

Actually, the cans won't be identical. In all likelihood, the mass-produced can will be far better-made than the machined can. Not only will it be thousands of times cheaper, it will be lighter, with fewer defects and blemishes. Why? Because a machine shop with a short production run might have just a thousand dollars to spend on getting it right, while a factory might have a hundred times that budget for every custom tool. The factory will have both means and incentive to make its tools sophisticated, precise and durable.

Beverage cans may be disposably cheap, but mass-produced items don't have to be expendable. The aviation industry has long made passenger flight affordable by mass-producing airframes durable enough for decades of everyday use.

Eclipse Aviation is trying to take this approach to a new level with its new 5-seat jet. Its very high development cost includes the price of licensing and automating a patented process called friction stir welding, which joins metal smoothly along a seam at low temperature. Unlike traditional welding, which heats the metal, friction stir preserves its temper, or hardening properties. This technique also leaves far fewer stress points than riveting. By automating friction stir, Eclipse hopes to significantly reduce errors and the need for mid-process inspections. Setting up this and other processes that improve safety is expensive, but it could lead to substantial savings later on: Eclipse expects very large production runs of perhaps a thousand small jets per year.

SpaceX, a company started by PayPal Founder Elon Musk, has set out to build beverage-can rockets. Like the Eclipse jet, SpaceX models its Falcon rocket with extensive computer simulations and uses friction stir welding machinery to build it. The Falcon is expendable, but the company may recover and re-use the spent stages, should that prove economical. If SpaceX can get its rocket to work, it may find itself in a good position to build toward a passenger rating—if it will focus on the passenger market.

NASA, Passengers and Competition

For decades, private rocket start-ups have either failed outright or fallen appallingly short of expectations. NASA staffers nod knowingly: it's a tough "biz," they say. Only they and their large contractors have spent the money and accumulated the experience needed to do it right.

Really? Take another look at the safety statistics on page 54. NASA hasn't made space launch any safer or cheaper, despite half a century and hundreds of billions of dollars of effort. No one has done it right yet. Is passenger space travel really that hard?

No one knows for sure. NASA sure doesn't. After all, passenger safety has never been its mission.

From its very beginnings in the Cold War, NASA's job was to make America first in space and keep it there. The agency wasn't chartered, as the FAA was decades earlier, to promote safe air commerce. NASA had foes, not markets; destinations, not products. It had less than a decade to beat the Soviets to the Moon with battleship-era hardware and computers a million times less reliable than today's mobile phones. So the agency cut corners. Rather than taking the time to design fundamentally safe vehicles, NASA borrowed expendable ballistic missiles and adapted them for military test pilots.

Decades later, NASA still takes its competitive mission very seriously. Its managers are skilled political operators. Its outreach organization (which I worked with through much of the 1990s) is formidable and tentacular. The Space Pharaohs really believe they own the cosmos. In their hallway talk, NASA personnel consistently regard outsiders, especially private launch efforts, as threats. This creates an environment hostile any form of passenger commerce in space.

It doesn't help that new space companies routinely shoot themselves in the foot by attacking lucrative satellite launch markets already owned by Boeing, Lockheed Martin and Northrop Grumman. These three giants have built nearly all of the launch systems for NASA and the Department of Defense since the late 1950s; they know their customers intimately and work hard to protect their proprietary interests against outsiders. When alerted to competition, they have an impressive arsenal at their command. They can act as "independent" industry consultants with would-be investors. NASA routinely sole-sources the Big Three and condemns their competitors as "inexperienced."

New companies are like new species: they usually do best in unfilled niches rather than those already brimming with competitors. It's simply unwise to try to take a big chunk of the most lucrative market away from a big entrenched competitor. The Honda Motor Company didn't threaten Detroit when it opened for business in the United States in 1958. Its first engines went into scooters and motorcycles. Apple Computer didn't threaten IBM in 1977 by building mainframes. Instead, it found a whole new market for tiny home computers and soon IBM was scrambling to play catch-up. Netscape, on the other hand, announced that it would crush Microsoft—and failed spectacularly.

Big companies generally own far more patents than small companies. In corporate patent law, the game is to sue, but settle out of court by trading cross-licensing agreements on each other's patents. This turns large competitors into collaborators and protects their incumbency. Small competitors, lacking a huge arsenal of patents, aren't dealt into this game and must fight in court. The majority of patent cases settle in favor of the plaintiff, with the defendant not only barred thereafter from using the technology in question, but also responsible for treble "lost revenue" damages and court fees. Fighting in court is expensive. So is settling. Seasoned competitors wield this sort of litigation like a scalpel, giving the upstarts a business-ending "cashectomy."[94]

Even expired patents can threaten a new company. Recently, SpaceX chose to use the pintle injector in its rockets. Injectors, which mix fuel and oxidizer just prior to combustion, typically have hundreds of holes. A pintle uses a single needle-like element to reduce fabrication cost dramatically with a slight performance penalty.[95] Northrop Grumman, a company with $33 billion in assets,[96] sued claiming trade secret protection. SpaceX counter-sued, claiming illegal extension of patent protection.[97] Not only could Northrop Grumman out-spend SpaceX ten-fold in this battle, but all delays caused by the case worked in the incumbent's favor. SpaceX undoubtedly faced pressure to postpone launches while the case was pending and spend money on lawyers, not engineers. It's a tough "biz" indeed.

Not only should new players in this field avoid attacking established niches, they should strive for near-invisibility during development. Blue Origin has been following this strategy for several years: as of this writing, the company's phone number remains unlisted.

The Environmental Challenge

Reaching space requires enormous expenditures of energy, so it is not surprising that rocketry has environmental impacts. The space industry withstands environmental scrutiny only because it delivers just a few people to orbit every year. It could not grow to truly inclusive scales before the environmental effects of current spacecraft operations from launch to landing physically limit its progress.

Proliferation and Bombs

Long before passenger jets became weapons in the 9/11 attacks, it was clear that any spacecraft could bring even the most distant war straight to one's doorstep in a matter of minutes. Von Braun's V-2 took about five minutes to reach London;[23] rockets capable of achieving orbit can reach any point on Earth in less than 44 minutes. That's why the very first satellite, the Soviet Union's Sputnik, shocked the world in 1957. In the midst of the Cold War, no one doubted that nuclear bombs might follow soon. Thus it is not surprising that U.S. law regards rockets as munitions subject to stringent export controls. This limits the scope and pace of international collaborations to the present day.[98]

Rockets typically carry large amounts of explosive propellants. Early in a development program, it is not unusual for a rocket stage to fly wildly out of control. International law holds a rocket's nation of origin liable for any damages it would cause in a crash. For this reason, and to protect bystanders who come to watch a launch, space ports often require that large rockets carry a destruct charge. Even the Space Shuttle carries Range Safety Command Destruct Systems—basically remote-controlled bombs—on each of its boosters.

A commercial passenger craft would never carry a bomb—who would board it? The pressure to put bombs on spacecraft will ease only to the extent that the vehicles prove reliable. Early passenger spacecraft therefore should be thoroughly tested and small enough not to cause enormous damage should one fail or fall to misuse.

Noise, Earth Impacts, Debris and Toxics

Spacecraft are noisy. The roar of the engines during launch can be so intense that Japan's Aerospace Exploration Agency (JAXA) has agreed not to launch from its seaside facilities on Tanegashima Island during fishing season (March-August and October-January).[99] When the Space Shuttle returns from space, two sharp sonic booms herald its arrival; that's why its launch and landing facilities are far from major cities. Clearly, Los Angeles International Airport will never become a space port. But Mojave Airport, a supersonic aircraft test site two hours to the North by car, is grooming itself for that role.

Every rocket launched to date has dropped spent stages, cargo fairings or explosive bolts onto land or sea. Spent satellites in LEO are routinely de-orbited, and many older satellites had components that made it all the way to the ground. Although ground controllers try to aim the falling satellites for the open sea, numerous satellite components have come down on land, including at least two nuclear reactors.[100]

Modern low-orbiting satellites are designed to fragment during re-entry into pieces small enough to vaporize. Because the satellites are relatively small, the heavy metals and other toxics this releases into the air add up to a minis-cule fraction of existing pollution from industrial, electronic and automotive sources. Of course, re-usable passenger spacecraft would not normally disintegrate on re-entry, so they probably would not change this situation for the worse. It may even be possible to use passenger spacecraft to deliver, capture and service satellites, as the Space Shuttle was designed to do. This would reduce the adverse environmental impact of satellites. But to keep their flight profiles manageable, it may prove wiser not to saddle the first few generations of passenger space vehicles with this burdening requirement.

Toxic propellants abound in rocketry: hydrazine, nitric acid and hydrogen peroxide have been popular with missile builders since the 1950s because they don't need refrigeration or high-pressure storage. But they are so corrosive and toxic that they require special procedures for safe handling, clean-up, storage and disposal. They are intensely regulated and for good reason: hydrogen peroxide (H_2O_2) tends to catch fire spontaneously when spilled on clothing. Hydrazine and nitric acid are also known to spontaneously explode under a wide range of circumstances.

XCOR Aerospace, a rocket company in Mojave, California, touts its use of non-toxic fuels such as alcohol and methane. When I asked CEO Jeff Greason about this in 2001, he replied that XCOR chose these fuels not out of environmental concern, but because of an intense allergy to regulation and paperwork. Their fuels have a long history of safe handling procedures with minimal training and gear. For oxidizer, they use liquid oxygen (LOX).

By contrast, Blue Origin's draft Environmental Assessment for its West Texas launch site declares the company's intent to burn rocket-grade kerosene (a relatively benign fuel) with hydrogen peroxide.[77] It will be difficult and costly to handle this toxic, corrosive, carcinogenic, highly flammable oxidizer on a scale needed for passenger transport. Workers will need to wear environmental "bunny suits" and train extensively for handling spills and leaks.

As the effects of past and present environmental degradation become more widespread, industry will come under ever greater pressure and incentive to develop clean practices. In this environment, the present space industry will have a tough time growing unless it stops using toxic propellants and throwaway components.

Space Pollution: Orbital Debris

The problem with throwing anything away is that there really is no "away," especially not in Earth orbit. As it stands today, NASA maintains a catalog of

9,680 artificial satellites, rocket bodies and debris down to about 10 centimeters across.[101] Objects between one and 10 cm are too small to track, but large enough to cause severe damage when they collide with a spacecraft. The European Space Agency estimates that there are between 100,000 and 150,000 orbiting objects larger than a centimeter across. What is all this stuff? The catalogs include everything from abandoned satellite fragments to spent upper stages to nuclear reactor cores to a lens cover. Forty-three percent of it is fragmentary debris from explosions and collisions in space.[102]

A secondary hazard exists in the form of paint flecks, aluminum flecks from solid rocket motor exhaust and other fine particles. These don't show up on radar, but sand-blast everything in orbit. Optics, solar panels, and other sensitive surfaces take a beating over time.

Collisions in space are mostly catastrophic. If two objects have circular orbits at the same altitude, either they are in the same orbit or their orbits will cross. If they are in exactly the right position when they cross, they will collide. The strength of the collision will depend on the angle between their orbits. Any angle larger than about three degrees will bring them together at supersonic speeds, destroying one or both of them completely.

Of course, orbits aren't ever perfectly circular or even elliptical. Earth is both bumpy and not quite spherical; some parts of it have slightly more gravity than others, perturbing orbits into shapes that never quite close. The atmosphere also exerts a slight drag on lower orbits, in amounts that vary substantially with altitude, sunlight and solar magnetic activity over time. This is bad for two objects that want to avoid a collision. Even if their present orbits would mathematically never bring them together, the orbits themselves are continuously changing slightly in unpredictable ways. Satellite-spotters know this, checking for updated orbits every few days. It takes constant vigilance with ground-based radar to keep the Shuttle advised of orbital hazards, and Shuttles have dodged to avoid several collisions over the years.

Drag from Earth's tenuous outer atmosphere eventually clears out the orbital debris, but the higher the altitude, the slower this process works. A fleck of metal below about 200 kilometers won't last a day, while above 1,000 kilometers it will remain aloft for over a thousand years.[103]

Our activities in space already have magnified the impact hazard in LEO far above the threat from meteorites. As a consequence, Space Station planners currently prepare for the possibility of a serious collision every decade or so.

The risk of collision scales roughly with the *square* of the number of objects in orbit. If a million people visit space in small vehicles as dirty as the 4,100 launches to date,[102] the time between severe collisions will drop to less than two hours.

The space industry therefore needs a complete overhaul of current procedures to achieve zero-debris operations from launch through landing. This includes new approaches to staging (with its current profusion of explosive bolts and throw-away fairings), collision avoidance and spent satellite removal. If we want a meaningful future in space, rockets must leave nothing behind.

Rising to the Challenges

By now it should be clear that none of the world's so-called space programs is really serious about putting people into space. This is tragic, considering how much has been spent, how few people have been there, how many have died trying, and how many still want to go.

A space industry built for millions of passengers cannot even slightly resemble today's launch industry. At minimum, it would involve a new vehicle that would:

1. Be designed for a very narrow range of flight profiles. For example, a passenger craft might carry people to orbit and simply return to the ground within hours. At most, it might dock with a station in orbit. It should *not* also have to provide life-support for up to two weeks, carry science labs, capture and service satellites, etc.

2. Perform a thousand successful test flights within its main profile before opening for commercial operations. That's ten times the number of launches that the Space Shuttle has made.

3. Use automated launch, landing and docking systems. The early test program should not use human pilots due to the high psychological and financial risks involved. Even during commercial operations, there may be no valid role for pilots.

4. Return to the ground ready to re-launch within a few hours and repeat this process thousands of times with only occasional down-time for maintenance and overhauls.

5. Contain all fasteners, mechanisms and solid exhausts, so as to leave no debris on-orbit. For example, there should be no rivets, bolts or screws exposed on the outer surface.

6. Use non-toxic liquid propellants to minimize regulatory red tape, legal liability, handling expenses and environmental damage.

7. Use extensive up-front study and routine monitoring to detect and prevent any possible damage to the Earth's atmosphere and magnetosphere.

8. Have investors willing to shun lucrative but entrenched satellite launch markets, ignore the opinions anyone employed by NASA or its contractors and focus instead on an untried new market (passenger travel).

There are no short-cuts. Failing to meet even one these requirements would fatally jeopardize either the spacecraft, widespread access to space, the orbital environment or Earth's environment. Remember, we're not talking about a few brave heroes and even fewer isolated launches. Anything done right or wrong would be multiplied a million-fold in a few short years.

NASA has never made any item on this list a priority because it is not structured to do so. Putting you and me and millions of others into space is a fundamentally inclusive proposition, while the Agency's mission is to establish American superiority in space—a competitive mandate that has nothing to do with commerce. To build a robust passenger space industry, we must renounce all but a tiny fraction of NASA's Cold War heritage, riddled as it is with crash efforts and short-cuts, and start over.

Sub-Orbital Development

On May 5, 1961, Alan Shepard rode the Mercury-Redstone 3 rocket to an altitude of 187 km and into the history books as America's first astronaut. Although he cleared the atmosphere, his vehicle never came close to reaching orbit. He was on the ground 15 minutes after launch.

Spacers have an easy time convincing each other that sub-orbital hops similar to Shepard's or *SpaceShipOne*'s will lay the groundwork for orbital craft and help pay for them, too. That's a tall order: reaching orbit, as we saw at the beginning of this chapter, is much harder than merely clearing the atmosphere. Virgin Galactic has quoted prices between $100,000 and $200,000 for the Shepard ride—comparable to the ticket prices ultimately possible for a trip to orbit (page 78). Surveys suggest that many people would save for a ticket to orbit, which would provide 87 minutes in freefall for every lap around Earth. But it's not at all clear how many people would pay just as much for only 3-5 minutes in space.

Similarly, the new sport of piloted rocket racing promises to use massive attendance and advertising revenues to advance the state of reliable, passenger-rated rockets. However, vehicles optimized for stadium attendance (low altitude, fast turns) have little in common with anything meant to survive a round-trip to orbit. Like their progenitor—the XCOR EZ-Rocket—these racers would have impressively large propellant tanks. But compared with spacecraft, their tanks are puny, providing a propellant mass fraction closer to 50% instead of the 90–95% required to reach orbit.[104] You also won't find heat shields and air-tight cabins aboard these low-flying crowd-pleasers.

While rocket races, Shepard rides and the like probably won't lead organically to orbital craft, they may help passenger space travel in two ways. First, they have the potential to show that rockets can be far safer and more reliable than anything NASA and its big contractors have produced to date. Second, they could demonstrate to investors that significant and diverse markets exist for passenger rockets. These would be two small steps forward.

Reaching Orbit

Nobody knows yet what safe, reliable transport to orbit might look like, though schemes abound: rockets launched in stages; rockets launched from airplanes, balloons or electric rails; air-fueled rockets; jets that become rockets as they leave the atmosphere; ultra-high altitude balloons with solar-powered ion engines that would take days to reach orbit; space elevators that take a month to climb giant cables all the way out to GEO.

Some approaches require fundamental technologies that haven't been invented yet. For example, the 40,000 kilometer cables that space elevators would ride require materials dozens of times stronger than anything yet made. Solar powered balloons require solar panels hundreds of times thinner and lighter than any built to date. No one has built and flown a jet-rocket with anything like the required performance.

Many schemes suffer from logistical problems. Space elevators are on a collision course with every satellite or piece of space debris that doesn't also orbit in resonance with Earth's rotation. Rockets launched from balloons can't launch on days when high-altitude winds blow strong. Multi-stage vehicles require separate recovery for each piece, often from distances of hundreds to thousands of kilometers apart.

Several schemes suffer from problems of scale: Electric rails would have to be hundreds of kilometers long to accelerate a vehicle to orbital speeds. They also would have to release the spacecraft at tens of kilometers altitude, a tall order, given that the Teipei 101, the world's biggest skyscraper, rises only half a kilometer above the ground. Vehicles launched from aircraft or even runways don't scale well to larger sizes, and wings add enormously to a spacecraft's weight.

From a financial perspective, anything new represents unknown risks and costs. So do complicated logistics. The space elevator falls down mightily on both counts, plus a third: it would cost about $16 billion before it could open for business—if everything goes as planned!

If passenger spacecraft design has a Prime Directive, it is to minimize the complexity of the system as a whole. Surprisingly, multistage rockets fare poorly by this criterion because each stage involves a separate set of rocket engines, tanks, pumps and landing systems. During flight, the stages must separate cleanly at supersonic speeds. After flight, they end up thousands of kilometers apart. To be re-usable (as is necessary for economical operations), they must be recovered, serviced and integrated (stacked together) again prior to launch.

Historically, rockets with some number of stages (call it N) went through less than N times as many tests as a single stage would. The stages of a Saturn V, for example, were stacked, tested and launched all together as a complete system. This scheme, called all-up testing, was invented not by von Braun (who was ever the incrementalist) but by George Mueller, who was made Head of Manned Spaceflight for NASA in 1963.[32] All-up testing was never about safety, though. It was an expedient designed to beat the Russians to the Moon. Safe passenger travel requires a hyper-von Braunian approach: N stages must be tested both separately and together, so they would need much more than N tests. That should scare investors. Nevertheless, all of today's launchers use multiple stages.

A Modest Proposal

Passenger space travel demands routine, economical operation. The ideal space vehicle would be as easy to use as a modern passenger jet: after loading people and propellants, it would swiftly reach orbit, circle the globe a few times, return safely to its scheduled spaceport and be ready to launch again within several hours. By far the simplest way to accomplish this is with a single-stage, rocket-powered vehicle.

All of the major aerospace companies have worked on Single Stage to Orbit (SSTO) designs at one time or another since the early 1960s. SSTOs promise better logistics (no stage recovery or mid-air refueling) and streamlined development: nothing lends itself to all-up testing like a single-stage vehicle.

For all their allure, SSTOs have been known to make astronautical engineers howl with rage. Most professionals in this field view such vehicles as uneconomical if not impossible. The problem with single-stage schemes is that propellants account for so much of their take-off mass that there's little room left in the design for anything else. To put this into perspective, when a 747 jet leaves the ground, 49% of its mass is fuel. Even Rutan's notoriously squirrelly *SpaceShipOne* took off with a propellant fraction of only 66%. By contrast, as we saw on page 50, a kerosene-LOX rocket requires a propellant fraction of 95% to reach orbit. That leaves too little mass for engines, pumps, tanks, heat shielding, landing gear and cargo. For over four decades, experts have looked at these numbers and just tossed up their hands.

It's possible to drop the propellant fraction to a much more reasonable 90% using hydrogen instead of kerosene, but the difference comes at a price. Hydrogen is eleven times less dense than kerosene, so it requires much larger storage tanks and piping. Because hydrogen boils at 20 K, it also requires insulation (kerosene doesn't). Hydrogen must even be kept insulated from the oxygen, which boils at a relatively balmy 90 K. The extra tankage, plumbing and insulation takes a serious bite out of the remaining 10% of the total mass at launch. Again, engineers acquainted with these numbers end up shaking their heads. Designing an SSTO is hard. Even a two-stage vehicle would drop the propellant fraction to a much more workable 80%.

The current situation with SSTOs resembles what Charles Lindbergh faced just prior to his famous flight across the Atlantic in 1927. Back then, hardly anyone believed that airplanes could cross oceans except in stages, stopping frequently to refuel as the crew of the NC-4 seaplane had done in 1919. The problem for aircraft of that era was much the same as it is now for spacecraft: the fuel fractions seemed prohibitively high for the materials then available.

Are propellant fractions above 90% that hard to attain? Not really: the late G. Harry Stine published a list (originally compiled by Major Mitchell Burnside Clapp) of 16 proven rocket stages that exceeded this criterion, some by as much as 6%.[105] The list includes the first and second stages of the Saturn V vehicle that carried astronauts to the Moon. The Saturn's second stage (designated S-II) used hydrogen-LOX propellants to achieve a propellant fraction of 91%.[106] Since it would only need a propellant fraction of 90% to reach orbit from the ground, it could have carried up to 1% of its launch mass as cargo.

That's the theory. None of these stages ever flew as an SSTO, and upgrading them would pose problems. The Saturn stages were expendable, so the S-II carried no heat shield or landing gear. Its engines fired just once, for 390 seconds, then fell back to Earth, crashing into the ocean with the rest of the stage. Upgrading its engines for routine use and adding durable re-entry and landing systems could easily increase the dry mass fraction by 1–2%. That leaves no room for cargo and probably puts us over budget for reaching orbit.

It gets worse. The S-II was made for use above much of the atmosphere, where the rocket's main job was to gain speed parallel to the ground. Its engines were sized to the task and therefore slightly too small to efficiently lift it directly against gravity from the launch site.[106] An SSTO would need 30% larger engines, which would add another half-percent to the dry mass fraction. The rockets also had long nozzles, optimized to work in vacuum. Short nozzles work better in the atmosphere. If an SSTO used long nozzles from the ground to orbit, the exhaust speed might suffer by as much as 10%. Two technologies have been proposed to get around this limitation: nozzles that lengthen on the way to orbit, and "aerospike" nozzles that use air pressure rather than the traditional bell-shaped chamber to shape the exhaust. Neither of these methods has made it into production yet, so for now let's just accept lower engine performance. That could easily increase the propellant fraction by 1%: now our S-II's dry mass is definitely over budget. There is no way it can reach orbit.

The S-II, however, was built heavier than necessary to function as an SSTO. Its total mass came to 449 tonnes fully loaded with propellant. Configured as an SSTO, its cargo might account for 1% of this: about 4.5 tonnes. But the S-II actually carried the third Saturn stage and the Apollo spacecraft: a total of 239 tonnes. That's over 50 times as much as it would need to carry as an SSTO.

A factor of fifty gives us some working room: with much lower structural loads, we could build an SSTO significantly lighter than an S-II without even using modern materials. However, we also have a much larger bag of mass-saving tricks to draw on than North American Aviation did when it built the Saturn stage forty years ago. Today's fiberglass composites are ten times stronger than the S-II's stainless steel structures. High-strength, heat-resistant plastics such as Magellan/Du Pont's M5 not only save mass, but are easy to form and bind with adhesives. New ceramics such as zirconium diboride are stronger and more heat-resistant than the carbon-carbon panels on the leading edge of the Shuttle's wings. Advanced materials and lighter construction can go far toward offsetting the mass penalties of durable re-entry, landing and engine systems. A dry mass of 8% may well be possible, leaving as much as 2% for cargo.

What would an SSTO look like? For starters, it probably would not have wings, as these add significant mass and complexity. Yes, the Shuttle has them, but it's also built to evade missiles during re-entry. Our passenger vehicle should be able to get by without this capability. It could use parachutes to land, but for daily operations, it will need something more precise and reliable. Perhaps it could take off and land vertically under rocket power.

This isn't as absurd as may seem. With nearly empty tanks, a hydrogen-LOX rocket would have a very low density with a correspondingly low terminal velocity: it only needs to fire rockets for about 6 seconds to come to a halt. Propellant lasts much longer during landing than at launch because the vehicle weighs 10 times less at the end of its journey. Thus it could hover for up to a minute on landing while consuming no more propellant than it did during the first five seconds of launch. That gives it a safety factor of 9.

Still, rocket-powered descent would be scary. The rockets and autopilot would have to prove themselves utterly reliable over many hundreds of test flights before any sane person would dare to climb aboard.

How would such a vehicle perform financially? Let's design it in some detail from the perspective of an investor, who will ask: What will our spacecraft cost to build, test and operate? How many people must buy tickets for it to break even in a reasonable time? What's the ticket price and what fraction of that amount is profit?

This last number, profit per ticket, or contribution margin, is an important business parameter. A high-risk investor won't even look at a margin below

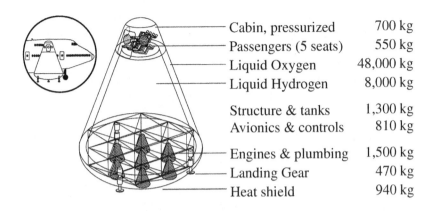

Cabin, pressurized	700 kg
Passengers (5 seats)	550 kg
Liquid Oxygen	48,000 kg
Liquid Hydrogen	8,000 kg
Structure & tanks	1,300 kg
Avionics & controls	810 kg
Engines & plumbing	1,500 kg
Landing Gear	470 kg
Heat shield	940 kg

Figure 2.1: Rough layout and mass breakdown for an 8×10-meter passenger SSTO with concentric tanks and a take-off mass of 62,500 kg. Inset (left) shows the vehicle with the nose of a 747–400 airliner for scale.

50%. Let's go with 75%, which means that a ticket will cost four times our operating and capital costs per passenger.

We also want something that will reach a financial break-even quickly. This has immediate implications for size and frequency of flights. If we build our vehicle too big, it may end up flying with empty seats or less frequently than it could. Nobody knows for sure how big the passenger market will be, so let's make our vehicle *small*. Let's see if we can fit five passengers, together with interior furnishings, pressure bulkheads and life-support into a mass budget of only 1,250 kilograms. That gives our vehicle a dry mass of only 5,700 kg including the passenger cabin: truly human-scale (figure 2.1).

Smaller means less development cost—an assumption that flies in the face of popular space folklore. Robert Truax has noted that the development cost of boosters historically has not correlated well with mass.[107] Even in aviation, cost tends to depend much more on the complexity of the vehicle, its expected use, the size of the test facilities, engineering labor costs, documentation requirements and the expected length of the production run than on the raw size of a vehicle. All true—within a factor of three or four. However, referring back to Table 2.4 (page 61), it's clear that the large airliners cost 30 to 40 times as much to develop as the tiny, ultra-modern Eclipse. Mass does matter, especially in a program that does sufficient testing. Even propellant and spares add up over a thousand flights.

Although we would spend more of our budget on testing than the Space Shuttle's builders did, we can save substantially in other areas. We would not have to develop, stack and recover multiple stages. Our vehicle also would have far fewer flight profiles than the Shuttle: no satellite servicing and delivery, no clever military maneuvers and no long stays on orbit. Automated launch and landing systems would save substantially on operations costs during testing and beyond. A passenger flight might have a safety officer/tour guide, but never a pilot. The total ground crew for flight monitoring, maintenance and fueling might amount to ten full-time staff per vehicle.

Let's budget twice as much per tonne as Eclipse Aviation spent developing its air taxi. At $400 million, our spending per tonne comes to 2.6 times as much as the Space Shuttle, 6 times as much as the 747 jet, and 40 times as much as *SpaceShipOne*. With a dry mass of 5,700 kg, our vehicle would cost $2.3 billion to develop.

How does development cost relate to the purchase price of a production vehicle? Clearly, development includes not only the cost of building the first vehicle, but also many one-time expenses such as design, tooling and testing. Boeing spent $11 billion developing the 747 airliner. As of 2006, the company had delivered 1,378 airframes.[108] When I plug the production run and development cost into the learning curve formula, I get about $2 billion per airframe.[93] But Boeing will sell you a 747 today for only $250 million.[80] So as a rough estimate of our space vehicle's retail price, I apply the learning curve to its development cost and divide by eight.

Let's build this vehicle to fly daily for 20 years, with 10-day overhauls at 3% of its purchase price every 50 flights. Suppose the owner's costs include staff salaries ($100,000 average per employee per year). Suppose further that making the engines reusable decreases their performance. For reference, the S-II's engines had a thrust-to-weight ratio of 66 and an exhaust speed of 4.13 km/s.[106] The Shuttle's Main Engine numbers are 55 and 4.44 km/s, respectively. Our vehicle's engine figures only need to be 50 and 3.97 km/s.

Our SSTO would be much squatter than a ballistic missile in order to reduce tank mass and improve its overall strength and handling. A wider shape does increase its exposure to atmospheric drag, but with a lift-off acceleration of only 1.2 gees, the vehicle would clear much of the atmosphere before reaching high speeds: the drag loss comes to only about 100 m/s.[109] However, fighting gravity during the slow ascent would cost 800 m/s. Fortunately, we have plenty of margin at this point: the ΔV to reach LEO quoted on page 50 already includes all maneuvering, gravity and drag losses our vehicle would experience, while ignoring the favorable 200-300 m/s boost Earth's rotation would provide when launching East.

The formulas to convert these demanding inputs into physical and financial performance figures are standard but numerous, so I have posted them on gaiome.com. Here's the bottom line:

A production run of 100 five-seaters able to reach a 290 km orbit would sell for $97 million apiece. With 300 full flights per vehicle per year, a rocket operator could break even in 6 months. Each passenger would pay $140,000 for a round-trip ticket to LEO (150 times less than Dennis Tito did). Cargo would fly for $550 per kilogram—90 times less than today's prices.

While $2.3 billion may seem like a lot to develop such a tiny craft, it's a bargain compared to the $33 billion it took to develop the Space Shuttle. Our vehicle may be small, but with 300 launches per year, each one would put an impressive 1,500 people into space annually—more than three times as many astronauts as have flown during the first fifty years of the Space Age.

Is it a good investment? I will leave that judgment to those with the deep pockets. Counting wealthy individuals, corporations and non-profits, thousands of separate non-government entities have the means to write the check today.

The small, automated LOX-H_2 SSTO is our best chance to meet all eight criteria for passenger spaceflight. With a propellant load only 2% as much as a jumbo jet, it should be a cinch to approve for routine, airplane-like operations. From production to operations to accident recovery, this vehicle would have far less environmental impact than today's rockets, leaving the way open for massive activity in space.

Habitats

With the planned 100 vehicles in service, the scenario above would put over 150,000 people into orbit every year—a $21 billion industry. Passengers would ascend to orbit in ten minutes, circle the world once or twice, and land a few hours later. For many, though, that's too little time.

The Earth is astonishing to watch from space. When people wandered into the NASA image facility I managed in the 1990s, most of them could not have cared less about our millions of close-up photos and maps of the other planets. No, they wanted to see our relatively meager collection of pictures and videos of Earth from space. Grainy as these images were, people came, day after day, to browse through them for hours.

According to those who have been to space, the real thing is much more captivating than any picture could express. From orbit, the Sun rises and sets every 90 minutes. The eye easily picks out systems of storms, ocean currents, forestry, agriculture, pollution and recovery. At night, city glow, aurorae and lightning speckle the dark globe. A living planet is so mesmerizing that it easily lures astronauts away from the clockwork tasks of science and survival.

Earth is the *only* living planet for all we know, and people will want more than a few hours to witness it. As demand rises for longer-term space habitats, they will be built.

But there's a problem. Our passenger SSTO isn't built for hauling large habitats to orbit. At most, we could modify it to carry a tonne or so of cargo, while serious space habitats can mass a hundred tonnes or more.

There are three paths to bridging this gap. The first is to build larger rockets capable of hauling the habitats up in one or two trips. Large boosters do exist today, but they're dirty, strewing spent stages, solid propellant, bolts, fairings and other debris everywhere. To avoid soiling their nests, space habitats need launchers as clean as our passenger SSTO. Simply scaling up from the passenger design would be possible, but a vehicle that could lift a hundred times as much would cost ten to a hundred billion dollars to develop: far too much.

The second approach would be to haul up the components for a habitat and assemble them on-orbit. There is no reason to think that traditional techniques involving robot arms, snap-together modules and astronauts in space suits would be any cheaper or easier than the ISS.

The third path is more interesting. What if we could haul up industrial feedstock such as metal wire, glass fiber and epoxy resins and use them to *manufacture* any desired structure on-orbit? Clearly, these techniques aren't mature, even on the ground. But for reasons that will become clear in Chapter 6, this sort of technology would have enormous long-term benefits, both in space and on Earth. Using this third path, space habitats would not have to await the development of very large, clean boosters. Modified passenger vehicles would do.

Even so, using the same development and launch costs per tonne as our rockets, a 110-tonne space habitat would cost some $10 billion to design, build and test, and another $58 million to deliver to orbit. But at $120,000 per room per night, it could break even in 18 months (we'll see this design again on page 189).

Early habitats of this type may be great places to visit and playfully test simple biological life-support systems, but they won't yet be worlds in their own right. They are far too small to support the diversity of dry land species that humans depend on for survival. Long before we can build enclosed ecologies that can sustain us indefinitely, we must learn more about how worlds work.

Chapter 3

Gaia and Her Children

"Nature is trying very hard to make us succeed, but nature does
not depend on us. We are not the only experiment."
—R. Buckminster Fuller[110]

Closed ecology experiments come in all sizes, though three in particular
stand out: (1) Gaia, the global biosphere, which has lived continuously for 3.8
billion years; (2) sealed terraria, popular now for the past century and a half;
and (3) semi-autonomous, semi-natural habitats where people have lived with
varying degrees of success for thousands of years (the modern incarnation of
which is called permaculture). Let's consider what each scale of experiment
has to teach us about world-building.

Gaia

Earth stands out among the planets as a living world, its atmosphere held far out
of chemical equilibrium primarily by plant life. If life and climate sustain each
other, together they resemble a sort of super-organism, which James Lovelock
has called Gaia.

The Gaia theory treats the entire Earth as a single living system—if not a
global organism, then at least a global metabolism. Plants, animals and microbes act together with global climate and geology to maintain the relatively
constant environment required for continued life. Just as an organism's metabolism maintains its temperature, pH levels, gas exchange, nutrient flow and

other parameters within definite limits, so Gaia maintains a whole host of parameters at the relatively constant levels necessary for life.

For example, the system of all Earth's plants and animals maintains oxygen at a level of 21% of the atmosphere, and CO_2 at 0.036%. If CO_2 levels rise, forests compensate to some degree by growing faster,[39] thus removing carbon from the atmosphere at a greater rate. In the absence overwhelming pressures (such as encroachment by human agriculture and settlement), this helps to maintain a stable climate.

The process of maintaining constant living parameters is called *homeostasis*. It is an emergent property of life, regulated not by a central organ but by active feedbacks among all of the world's components—including the nonliving ones. In the case of respiration, plants require carbon dioxide, animals require oxygen, and the atmosphere mediates the ratio of the two gasses. Other interactions, such as those involved in maintaining global temperatures and circulating nutrients, are more complex.

Origin in Planetary Program

The first hints of Gaia theory appeared in the early 1960s when biologist James Lovelock was working on the Viking missions to Mars. One of his jobs was to design experiments intended to detect life on Red Planet. He felt that most experiments proposed by mission engineers were too specific to Earth life—for example, preparing a "soup" that Earth microbes would eat, then watching for chemical changes that would indicate it was being consumed, metabolized, or respired.

Lovelock took a wider view. He and colleague Dian Hitchcock noticed that the atmosphere of Mars—mostly CO_2—is in chemical equilibrium with itself and the Martian soils. By contrast, Earth's atmosphere is wildly out of equilibrium: without life, it could not sustain its present mix of gases.[111] Oxygen in particular is far too reactive to remain in its present abundance in the atmosphere without an active source. It was patently obvious to Lovelock and Hitchcock that life maintained this disequilibrium. In their view, Earth, with an atmosphere rich in chemical potential energy, stood out among the planets as a living world. Mars, with its thin atmosphere so close to equilibrium, was hopelessly dead.

Thinking of the striking difference between muted Mars and the blue-green jewel of Earth, Lovelock came to the view that life "could not hang on in a few oases, except at the beginning or end of its tenure . . . unless organisms occupy their planet extensively and evolve with it as a single system, the conditions of their tenancy are not met. The system of organisms and their planet, Gaia for short, must be able to regulate its climate and chemical state." [112]

Invariants

The claim that Earth is alive garnered enormous attention, of course. But was it actually a theory?

Theories—mathematical models in particular—are the actionable parts of science. Without them, we're left either with pretty but useless notions or mere measurements with no predictive power. What we seek are the invariants: precise myths that are truer than the truth we've known thus far, that shed light not only on the past and what we've found, but also on the future, on worlds of possibility not yet discovered, dreamed or made. Einstein's energy equation ($E = MC^2$) is great theory not merely because it explained the phenomena of his time, but because it helps us understand thousands of things—from the CCD chips in digital cameras to quasars at the edge of the universe—that Einstein never observed or even dreamed of in his lifetime.

Finding invariants is hard in biology because life is enormously more complex than the simple solids and statistical processes of physics. Many important phenomena, such as metabolism, emerge from the interaction of tremendously large and complex systems whose operation does not easily reduce to the mere exposition of their working elements. For example, if you grind up a live sponge in a blender and return it to its environment, it will reassemble itself. Although it is possible to describe the process by which it can do this, no one has found an invariant that would confer this ability to other organisms such as a mouse or a person. Looking for physical and chemical invariants beyond the obvious (such as "total energy into a system = total energy out") can be confounding in the biological world, like looking for a needle in a haystack.

To penetrate this fog of complexity, many have found it fruitful to turn away from the literal world and construct minimally detailed myths. Thus, as the Gaia theory came under fire for its initial dearth of invariants, Lovelock invented a myth that distilled Gaia's complex processes to their very essence.

Daisyworld

Imagine a hypothetical planet populated exclusively by two or more varieties of daisy. The daisies are either white or black, or (in some models) varying shades of gray. The soil is equally fertile everywhere. Daisyworld lacks air, water, and all other environmental variables except one: temperature. The daisies die if the temperature is too cold ($10\,^{\circ}$C) or too hot ($30\,^{\circ}$C), and reproduce fastest near the middle of this range ($20\,^{\circ}$C).

Daisyworld has no clouds, seas or greenhouse gases to complicate the model; the temperature of the surface depends on just two things: the amount of sunlight received, and the albedo, i.e. the fraction of sunlight that it reflects back into space. Daisies range from an albedo of 0 (black) to 1 (white). Just as black cars absorb sunlight better than white ones and hence get hotter on a summer day, so lower albedo daisies would warm their immediate surroundings more than their higher-albedo counterparts.

Daisyworld orbits a star that is getting brighter. In the absence of daisies, Daisyworld would simply heat up over time.

If the daisies got started when the temperature was $10\,^{\circ}$C planet-wide, the black daisies would soon dominate, because wherever they abound, they would absorb sunlight and raise the local temperature, making the land there more conducive to the growth of all daisies. This would soon lead to a global preponderance of black daisies, which would lower the planet's albedo. As a result, the global average temperature would rise above what it would have been as a gray, "dead" planet.

If the daisies completely controlled the planet's albedo, the average temperature of Daisyworld would very quickly hit $20\,^{\circ}$C, after a brief overshoot to $25\,^{\circ}$C.

As the star brightens, the black daisies enjoy less and less of an advantage; more white daisies start to appear. When the star brightens to the point that the "dead" temperature of Daisyworld would have averaged $20\,^{\circ}$C, the populations balance. As the star brightens, the black daisy population dwindles, and the white daisies begin to dominate. The planet's total albedo climbs, maintaining a lower surface temperature than it would have had as a dead planet.

As it happens, the Sun has brightened some 30% over the course of Earth's history, yet global average temperatures have remained relatively constant.

Critics had called the Gaia theory teleological, meaning that it seemed to require that plants and animals actively coordinate their growth and behavior with the conscious goal of maintaining a stable environment. Lovelock and Watson invented Daisyworld specifically to head off this criticism. The daisies have no ability to choose at all: they simply can't survive outside of a certain temperature range, and each daisy either raises or lowers the temperature of its immediate environment slightly depending on its albedo. Gaia emerges in Daisyworld not as a choice the daisies are able to make, but as an automatic interaction between them and their environment.

Daisyworld has its shortcomings. It ignores the process of evolution, despite the presence of a strong environmental selection pressure. Some have argued that the model itself is teleological because it chooses daisies that could influence the ability of their environment to sustain them, yet assigns them identical temperature sensitivity.[113]

Nevertheless, Daisyworld does strip Gaia's mechanics down to its barest mathematical essence. At this level, Daisyworld bears some similarity to the work of Lotka and Volterra, who modeled populations of predators and prey such as wolves and rabbits. The Lotka-Volterra equations succeeded in predicting stable populations as well as plagues and famines. But they also confounded theorists by becoming less stable as additional species were added. This is a very non-intuitive result. It is hard to imagine that a forest brimming with thousands of species is less robust than the monoculture of a tree farm.

What could be missing in the Lotka-Volterra equations? Well, they neglect the environment and any contributions or susceptibility that each species might have to it. But these interactions are exactly what Daisyworld models.

The first Daisyworlds ignored species diversity and predator-prey relationships. When Lovelock added these to his model (e.g. daisy-eating rabbits, rabbit-eating foxes, and a wider range of gray daisy species), he found that the stability of temperatures and populations improved as the number of species increased. So did recovery times from severe perturbations such as plagues.[112] As a correction to Lotka-Volterra, Daisyworld was a huge success, correctly modeling the observed tendency for environments with greater species diversity to be more stable.

Whether or not Gaia stands the test of time as a profoundly useful theory of planet Earth, it is at the very least a helpful way to think about the design of living worlds—especially artificial ones. Therefore, I shall treat it as "true" from here on.

To minimize the work needed to survive in a gaiome, one of our design goals will be to get as many physical and biological parameters as possible to self-regulate. These might include atmospheric temperature and pressure, the partial pressures of oxygen, carbon dioxide and other gases, fixation of carbon and nitrogen, soil pH and nutrient balance, and aquatic salinity. Taking cues from Gaia, we should recruit appropriate species of living beings for these tasks at every opportunity. Unlike machines, though, live organisms can maintain themselves and establish or repair their links to each other—and to us. When we introduce species, we would be wise to remember that they are really independent partners with their own agendas and much too sophisticated to treat as cogs.

Universal Enzymes

Homeostasis has long been observed throughout entire *biota*, or classes of organisms that share a common environment. In 1934, for example, Alfred Redfield noted a constant relationship between carbon, nitrogen, and phosphorus in marine organisms. Living cells maintain these ratios internally by regulating how they process nutrients from their environment. The work usually falls to enzymes that have evolved in synergistic clusters where each enhances the function of the other.[113]

Does Gaia herself have enzymes, sprinkled among the diverse species of the world, that ensure homeostasis? Organisms do worse when their environment lacks vital nutrients, so it's plausible that their enzymes evolved in tandem with the Gaian feedbacks of global climate. Biologist George Williams has described the workings of three such enzymes in the nitrogen cycle, and showed how these might provide the needed feedbacks throughout the biosphere. He suggests that further studies along these lines might reduce Gaia to an exact and detailed science, just as biochemistry is starting to make sense of physiology.[113]

Until then, our design work must resort to experience, intuition, and broad indicators of ecosystem health. One such indicator is the cycling ratio.

Cycling Ratios

If life used its primary elements—carbon, hydrogen, oxygen, nitrogen, phosphorus and sulfur—just once and permanently disposed of them, it would not last long. Gaia recycles.[114]

A volcano belches forth an atom of carbon, probably bound to an oxygen atom as CO_2. Over time, that carbon atom will become part of some 200 living beings—inhaled by plants, eaten and exhaled by animals and decomposers, borne by winds throughout the world, again and again—until it's eventually lost and buried as part of the calcium carbonate skeleton of a marine organism on the ocean floor. Other elements fare even better: nitrogen gets recycled up to 1,300 times before also being buried in the ocean floor.[39]

An ideal Gaian system would be materially closed, meaning nothing comes in or goes out except for energy in the form of sunlight. Indeed, Earth comes close to this ideal, exchanging a vanishingly tiny fraction of its mass with the rest of the solar system, and efficiently re-using many of the elements required for life.

But Gaia is not perfect. All of the elements of life are eventually buried or lost to space. Some elements, such as calcium, are so ubiquitous in rocks that life can apparently afford to waste it with abandon: an atom of calcium typically cycles through just one living organism before being lost again as a mineral. In other words, Gaia is slow to recycle teeth and inefficient at handling calcium runoff.

In designing gaiomes, we'll have the luxury of choosing groups of species to increase the cycling ratios of vital elements. For example, in tropical rainforests, calcium has a cycling ratio of 50, so wet tropical biomes might go a long way toward closing this cycle.[115] Even then, though, we'll need machinery to close the inevitable leaks and return elements back into circulation. But the higher the cycling ratios, the smaller these machines can be.

Reservoirs

Natural environments tend to store useful resources in reservoirs of various sizes. We can think of these as nature's savings accounts. For example, the atmosphere has a large reservoir of oxygen that all animals (including you and me) draw on. To get a sense of how big it is, if every oxygen-producing organism died today, it would take several million years for respiration, weathering and combustion to deplete the atmosphere's oxygen supply to half its current level.[113] Human activity over the next few centuries will not "choke" the planet, though our CO_2 emissions might very well raise the temperature to lethal levels in a short amount of time.

The air volume available to each inhabitant of most space habitat designs is thousands to millions of times less than on Earth. The simplest gaiome design in Chapter 5 is spherical, which from one perspective is very wasteful because it requires the most air per person. But this maximizes its reservoir of oxygen. If all the plants in even the smallest gaiome died, its oxygen would last about a hundred years. While that's much shorter than the millions of years we would have on Earth, it's still a lot of time to try growing again from seed.

Natural Capital

Earth may be rich in oxygen, but we're rapidly spending down other accounts. The store of economically accessible oil may not even last until mid-century.[116] The world food reserve already fails to meet world demand to the point where millions are already starving.[48] Should a major crop such as wheat or soy fail, that number could rise into the hundreds of millions. Worse, pollution and habitat loss are rapidly destroying the planet's natural capital.

Economic capital refers to income-producing assets such as factories, real estate and agriculture. Natural capital, by analogy, includes anything that produces biological wealth. For a long time now, back to the Oxygen Catastrophe, the dominant form of natural capital on the planet has been chloroplasts. These are the green organelles, found in leaf and algae cells, that steadily convert the Sun's energy into the sugars that power plant and animal life. Unfortunately, much of our economic wealth comes from using or destroying this natural capital much faster than it regenerates. In other words, we're paying for our amped-up lifestyle by liquidating Gaia's assets.[117] As we clear out forests, high-quality lumber grows scarce. As we pave over rich soil with polluting roads and industry, we destroy its fertility

Gaiomes would have small reservoirs, making them especially vulnerable to mismanaged natural capital. It would be unthinkable to dump semiconductor reagents into a gaiome's soil, as has happened in once-fertile Santa Clara Valley, not only because these chemicals are often toxic, but also because we would lose them as valuable resources in their own right.

The central ethic of ecology is conservation. A healthy ecosystem makes efficient use of available energy and moves nutrients through reciprocal loops with high cycling ratios. It achieves these ends through biomass and diversity. The sheer mass of plants and other chloroplast-bearing organisms determines how much of the energy and nutrients flowing into a region life can capture. The total number of species, in turn, determines how efficiently life can utilize these captured flows. For example, a desert and a rainforest may receive similar amounts of sunlight and rain, but with thousands of times more biodiversity, the rainforest has much more natural capital.

The late urban theorist Jane Jacobs has argued persuasively that economics is a special case of ecology. In her view, locally diverse economies generate wealth just like any other healthy ecology: by conserving and cycling resources among many participants. Thus internal factors have at least as much power to determine the wealth of a community as the external ones. [118] Gaiomes, with their small size, material closure and fixed solar budget, would put this notion to the ultimate test.

Biospheres

Humans have tried to control the climate of their dwellings for many thousands of years. As many of our ancestors found their way into cooler climates, not only did they learn to dress warmer and hunt and farm with the seasons, they also built their houses thicker and fitted them with hotter hearths.

But none of these challenges can compare with the difficulty of setting up a house in space. Our needs there go well beyond shelter and warmth: we would lack air, water, food, building materials and the plants that freely regenerate them for us on Earth, even in Alaska, Siberia and the other frigid "wastelands." We may even lack gravity. To settle in space, it is not enough to build a house. We must first build a world. As it happens, people have been trying to build miniature worlds for a long time.

Early Work

In 1895, when Tsiolkovsky pointed out the need for space travelers to build a sealed, regenerative environment, [60] his readers could readily grasp the concept. He was talking about a terrarium, an invention that had been popular for decades.

In 1829, Nathanial Ward, a London physician, had lost much of his garden of mosses and ferns to the city's air pollution. In desperation, he sealed a prized fern inside a bottle. It thrived, and Ward quickly expanded his experiments. [119] Wardian cases, as these early terraria were called, became popular in London in the 1830s. By the early 1840s, explorers such as Joseph Dalton Hooker began to use them to return living specimens from distant climates. By 1851, Ward displayed a case containing a fern and mosses that had not been watered in 18 years. In the 1850s and 1860's, Wardian cases became fashionable in Victorian households and in the United States.[120]

In 1926, Vladimir I. Vernadsky published *The Biosphere*, which suggested that earth life actively maintains the atmosphere in a state of disequilibrium favorable to its continued survival.[15] His work, along with Tsiolkovsky's, laid the foundation for biospheric studies in Russia.

Soviet Bios Experiments

When the Soviet Union began its cosmonautics program, visionary scientists immediately recognized the need to provide long-term life support to future explorers. Because Vernadsky's work was popular in the USSR, the notion of using closed biological systems for life suppport made sense to Russian scientists. So in 1960-1961, Yevgeny Shepelev and Gana I. Meleshko of the Institute of Aerospace Medicine in Moscow built a chamber that used photo-synthesizing *chlorella* (blue-green) algae to remove carbon dioxide from the air and provide oxygen. Although the experiment successfully sustained Shepelev for 24 hours, trace gases (presumably including hydrogen sulfide) had built up to alarming levels. When the chamber opened, colleagues were aghast at the intense odor.[121]

In 1965, researchers at the Department of Biophysics in the academic city of Krasnoyarsk expanded this work with a 12 cubic meter chamber (about the size of a large home lavatory) called Bios-1. This system used *chlorella* algae to recycle air, but not water or food.

In 1968, the team added a second chamber of about the same size for grow-ing higher plants such as wheat, beet-roots, carrots, cucumbers, and dill. The chamber quickly acquired the name *phytotron*, from a 1940s parody of the im-pressive cyclotrons found in physics. The phytotron's plants purified about 25% of the air, and the much smaller algae cultivator handled the remaining 75%. The combined facility went by the name of Bios-2.[122]

In 1972, this effort expanded massively with the construction of Bios-3, a fully sealed facility occupying $14 \times 9 \times 2.5$ meters—about the size of a large apartment or small bungalow. Bios-3 was divided into four equal compart-ments, with a crew of up to three people occupying one of them, phytotrons occupying two more. The experimenters eventually replaced the algae unit in the fourth compartment with a phytotron. Although the algae proved very easy to grow, it required enormous effort to process into palatable food, which itself proved nutritionally deficient. The higher plants required more tending, but did a much better job feeding the crew. Two phytotrons also proved more than sufficient to remove carbon dioxide from the air, replenishing it with oxygen.

A biospherics experiment is called a closure. Complete closure would imply 100% recycling of all air, nutrients and water (but not energy, which is always required by living systems). In the case of Bios-3, the food cycle was not completely closed. Meat was brought in from outside to supplement the food grown in the phytotrons. Human feces was not recycled but instead dried and stored for analysis. Catalytic converters removed some trace organic gases from the air. Under these conditions, Bios-3 ran three closures with crews of two to three people for six, five and four months, respectively. These helped researchers figure out how crop area and light requirements depend on the amount of closure—and how much weight phytotrons might save them on long-duration space missions.

The bottom line? Without recycling, a crew member might consume 13 kilograms of food, water and air per day. Using phytotrons, that number dropped to 0.6 kilograms per day. [122]

Ecospheres

Research elsewhere in the world around this time was a lot more modest. In 1963, Jack Myers at the University of Texas developed an experiment similar to Bios-1, using mice and chlorella algae. [123] In 1968, Professor Clair Folsome of the University of Hawaii sealed several samples of water from nearby beaches in half-filled air-tight glass spheres, and set them in indirect lighting in his office. The algae and bacteria in this 100% closure quickly stabilized with slightly higher oxygen content than Earth's atmosphere and lived on for decades, powered only by diffuse sunlight, without showing signs of degeneration. [124] It was a big surprise that a sample, randomly scooped from its environment, would spontaneously form a stable ecosystem.

As of 2005 you could still buy commercial derivatives of Folsome's work from vendors such as Stockley's Aquariums (www.ecosaqua.com), who will mail you a stable ecosystem of water, air, algae, brine shrimp and bacteria inside a sealed glass capsule. The organisms within them are supposed to survive about two years without additional water, air or nutrients. Some shrimp colonies have reportedly survived fifteen years or more, though none in the three ecospheres I bought for family members lasted longer than six months.

Allow Time for Assembly

Folsome and other experimenters showed that stable ecosystems are apt to emerge in any closure. However, building a stable habitat for any specific non-microbial species (such as humans!) requires significant effort, resources and time.

Ecosystems stabilize at different rates. Hardwood forest ecologies take decades to centuries to mature.[125] However, simpler systems, can become established much faster with human inputs. Efforts to restore dozens of hectares of tall-grass prairie in East-central Illinois have met with success in about 7 years.[126] It took 5 years for Steinhart Aquarium to stabilize its 75 cubic meter coral reef tank.[121,127] The Flowering Tree Permaculture Institute in New Mexico took a mere 4 years to transform a patch of barren high desert into a lush green oasis with over 500 species of plants.[115] So it may be possible to make a gaiome self-sustaining in just a few years.

In 1971, a research group called the New Alchemists set out to design *bioshelters*, i.e. human dwellings that would use indoor plantings to process sewage and wash-water providing, in the process, warmth and energy. These dwellings, however, were not entirely closed. Co-founder John Todd noted that The Arc, the group's bioshelter on Prince Edward Island, Canada, succeeded because its "gaseous, climatic and biological health is created through its couplings and linkages with the exterior environment."[3] He further pointed out that space habitats would lack these beneficial connections to Gaia. Indeed, the spontaneous arrival of unexpected species had helped close the ecological gaps in the Flowering Tree Permaculture.[115] From these experiences, we should plan to keep young gaiomes connected to Earth's vast biological reservoirs long enough to establish stable ecologies—a significant challenge when our only links are one-tonne rocket cargoes that cost half a million dollars per launch.

CELSS

In 1977, NASA began research on Controlled Ecological Life Support Systems (CELSS —in some papers, the "C" stands for "Closed"[39]). Under this program, university researchers such as Frank Salisbury and his team at Utah State University grew plants indoors in hydroponic trays (no soil, just nutrient-rich water) under artificial lights. The researchers reported a tenfold increase in yields over field agriculture with this approach. [124]

In 1985, NASA funded several facilities to test crops and water/solids recycling. Over the next two years, workers at the Kennedy Space Center in Florida converted a surplus hypobaric chamber into a cottage-sized, sealed facility. Like an electronics breadboard, this Biomass Production Chamber (BPC) let researchers "plug" their experiments into modular trays, controlling and measuring multiple variables such as nutrients, lighting and temperature. [128] The BPC could, in theory, supply all of a single inhabitant's air, water and food needs. More often, though, the system ran without anyone inside.

From 1988 to 2001, BPC experimenters grew numerous crops including wheat, soy, beets and strawberries under a wide range of conditions. From these tests, they determined that the amount of food they could grow in a given space depended on the amount of light the plants got. When they doubled the light, for example by making it brighter, by making it shine for more hours each day, or both, they found they could grow twice as much wheat. Of course, practical limits exist for each type of plant. Tomatoes and potatoes grow best with alternating periods of darkness while wheat prefers constant illumination and can take, overall, about twice as much light. [128] These were some of the findings I will use to estimate the size and lighting requirements for the gaiome's gardens in Chapter 5.

Biosphere 2

By far the most ambitious experiment in biospherics began its gradual genesis in 1973, when Texas oil and real estate heir Edward Bass, Jr. dropped out of Yale to join the Synergia Ranch commune just south of Santa Fe, New Mexico. There he met John Allen, a theater group leader, [129] along with Allen's close associates of six years, Margaret Augustine and Mark Nelson. [124]

Over the years, the four of them worked together on numerous ecology-oriented ventures. Their London *Institute of Ecotechnics* consulted on "integrating appropriate technology with ecological restoration projects." [123] In 1975, they launched the *RV Heraclitus*, an 84-foot Chinese Junk that has since sailed over 300,000 km and is currently documenting the health of coral reefs worldwide. They also opened sustainable organic agriculture projects in France

(*Les Marronniers*, 1975), Australia (*Birdwood Downs*, 1979) and Puerto Rico (*Las Casas de la Selva*, 1982). Like Synergia, several of these venues also function as conference and arts centers.

In 1982, *Les Marronniers* hosted the seventh of a series of Institute of Ecotechnics conferences that had been focusing on the ecology of various biomes, then of Earth as a whole. Titled *The Galactic Conference*, it featured invited speaker R. Buckminster Fuller, who angrily challenged the 80 attendees to start designing artificial biospheres. Fortunately, such a design existed: T. C. Hawes, who had met Ed Bass a decade earlier while installing plumbing in his house, had designed a basketball-sized architectural model of a spherical space ship that would allow humans to travel indefinitely through the cosmos.[130]

This tangible model galvanized the participants, who decided, after much additional discussion, that the time had come to model Earth's biomes, not individually, but all together as an interacting system that simulated the workings of Earth's entire biosphere.[124] Such a model might provide insights into what it would take to build off-world habitats, either free floating, as in Hawes' design, or on the surfaces of the Moon and Mars.

The details of their design, named Biosphere 2 (Earth being Biosphere 1), took shape at the Institute's *Biospheres Conference* in December, 1984. The goal of the conference was to design an experiment that would include all five of life's kindoms[131] in three to five biomes—an ark of sorts that could survive for centuries, preserving its sealed biomass from the radioactive fallout of nuclear disasters.[132] Conference organizers even saw Biosphere 2 as the next step in the evolutionary progress of life (a theme I'll revisit in Chapter 7). It was quickly decided that it should hold a crew of eight people, a stable social unit that Allen likened to a platoon.[124]

Thus far, only Folsome's tiny flasks had achieved full closure, and they encompassed a hundred million times less volume than the proposed Biosphere 2. So the Institute's newly-formed Space Biospheres Ventures (SBV) wisely chose to build a Test Module just large enough to enclose a single person: a cottage-sized structure about 1,000 times smaller than Biosphere 2, yet 100,000 times larger than Folsome's flask. Design work on the Test Module began in 1985.

The Tests

By January 1987, the Test Module was complete. Closure experiments commenced, using organisms from all the proposed biomes of Biosphere 2. Human closures began in September 1988. At the end of the first experiment, which lasted three days, the levels of oxygen and CO_2 were still fluctuating substantially. By the end of the third and final closure, lasting 21 days in November 1989, the oxygen and CO_2 levels stabilized. The system had achieved equilibrium.

Beyond the concept of a fully closed biosphere containing a human being, the Test Module also validated SBV's new construction methods. Biosphere 2 was meant as a prototype for biospheres that would last centuries, so it needed to achieve a leak rate much lower than 1% a year. By comparison, Soviet Bios and NASA CELSS experiments leaked anywhere from 1% to 10% per day.[123]

In practice, the Test Module far out-performed the Soviet and NASA experiments, yet it never achieved anything like its goal of 1% closure per year. Published estimates of its projected annual leak rate range from 8% (according to Institute founder John Allen[124]) to 24% (according to Mark Nelson,[123] one of the "bionauts" who would later be sealed inside Biosphere 2). The leakage could have been a serious issue for the builders, but they reasoned as follows: the volume of Biosphere 2 is 1,000 times larger than the Test Module, but its surface area (through which leaks occur) is only 100 times larger. Thus it should take $1,000 \div 100 = 10$ times longer for the same fraction of air to leak out of it—assuming the same quality of construction. SBV therefore expected Biosphere 2 to have an annual leak rate somewhere between 0.8% and 2.4%. As it turned out, it was closer to 10%.

There was something odd about the Test Module. While it apparently provided far more robust closure than any prior terrarium, SBV began construction of Biosphere 2 three years before the Test Module completed its closures. This made its lessons much more expensive to apply. Worse, SBV halted research with the Test Module as soon as it obtained a stable closure. Thus in one swift move, SBV threw out the fundamental tenet of scientific research: replicability.

Thomas Edison's light bulb reputedly took over 10,000 experiments to perfect. Drugs require extensive and repeated clinical trials for approval. Early attempts to grow coral reefs in aquariums took hundreds of trials before anyone succeeded. Theories of stellar evolution are constrained by hundreds of thousands of cataloged spectra. Eukaryotes took trillions of generations of perhaps 10^{31} individual organisms to evolve from prokaryotes. Through variation, experiments make iterative improvements in the underlying process or theory. Through repetition, experiments validate process or theory by precisely predicting or duplicating a result. The quality of research always boils down to the quality, number and repeatability of the experiments.

If the Test Module had been a science experiment, there would have been more than one of it. These would have run continuous closures for years at a time, perhaps rotating crews every couple of weeks to test the leak capacity of its airlocks. During their prolonged runs, multiple Test Modules could have improved and validated not only the engineering needed to operate Biosphere 2, but also its financial underpinnings in tours and technology licensing.

But SBV was rushing a deadline. In 1984, NASA announced plans to have Space Station *Freedom* in orbit by 1992. SBV wanted to develop and patent life support technologies before then and sell them to NASA and its tenants aboard *Freedom*.[124] Rather than derive these technologies from the successful Test Module (and possible replicates), SBV chose to throw all its research dollars—some 20 times the cost of the Test Module—into a single monolithic experiment. Like the Apollo Moon landings, Biosphere 2 would be "one giant leap:" an all-up test of a complete ecological microcosm of Gaia.

The Closure

Biosphere 2 ultimately housed seven biomes, including a 4 million-liter artificial "ocean," a rainforest, a fog desert, and residential/agricultural areas. Construction of the 3-acre glass-topped facility ended in 1991, and almost immediately it was sealed with a crew of four men and four women.

The first closure was intensely challenging. The Arizona site had been chosen for its relatively abundant sunshine, but as luck wouldn't have it, the closure coincided with El Niño, a then-uncommon weather pattern that brought extra clouds. This meant less sunlight, which meant less plant growth. The crew had planned a diet of 2,800 calories per person per day, but the 80 struggling crops in their agriculture unit only provided 1,750 calories of food. On this diet, the crew lost, on average, 13.5% of their body weight.[133,134]

The decreased photosynthesis meant less carbon was removed from the atmosphere. At times, CO_2 concentrations reached 4,000 parts per million, about twice as much as in a crowded city and over ten times as much as Earth's atmosphere overall.[121,133] While not life-threatening, these levels were worrisome. More worrisome was the oxygen level, which fell from 21% down to 14%.[124] The carbon dioxide buildup was far too small to explain the decline in oxygen. Where was the oxygen going?

It turned out that Biosphere 2's soils were about six times too rich in organic materials such as peat and compost. These had been added to the natural soil (which typically contains about 5% organics) to speed up plant growth. The microorganisms that make these soils nutritious for plants consume oxygen and exhale carbon dioxide. As the carbon dioxide built up, it reacted with the concrete, which had not been fully cured at closure. Whereas plants remove carbon and return oxygen to the atmosphere, the concrete removed carbon dioxide, but kept both the carbon and the oxygen.[135] Had the concrete been given time to cure, the CO_2 problem would have appeared much faster than it did. As it was, even prior to closure, the project noticed high CO_2 concentrations and quietly installed "scrubbers" similar to those found on submarines. These came as a surprise to the science community and general public when they were disclosed later in the closure. On December 20, 1991, less than three months after closure, 17,000 cubic meters of air (about 10% of the total volume) were imported to make up for air that had been lost.[136]

About this time, allegations began to surface that Allen's group was more of a militaristic cult than a true research institute. Former members claimed they were subjected to constant physical abuse, seven-day work weeks and lengthy harangues at mealtimes. They had little or no privacy and were actively separated from their siblings, parents, and young children.[136,137] A *Time* magazine article exposed the Institute of Ecotechnics as "an art gallery and cafe," casting serious doubts on the degrees it had conferred.[138] In particular, according to the article, Biosphere architect Margaret Augustine's only degree was from the Institute.

The loss of credibility hurt the project and scared scientific partners away. But the group's focus on control and secrecy ultimately harmed the project even more. In May of 1992, the Science Advisory Committee recommended that the data collected up to that point be opened to outside scrutiny. SBV disregarded this request, presumably because the data constituted trade secrets. The advisors had also cautioned against using rich soils, but were ignored. The crew inside the biosphere, suffocating from lack of oxygen (as a result of the soils), split into two equal factions over whether or not to bring in oxygen. The anti-oxygen faction was loyal to Allen and Augustine in Mission Control. Their theory was that by growing more plants and sequestering more wood (hence carbon) in the basement, they could resolve the crisis. They stopped speaking to the pro-oxygen faction, who were beginning to resent Mission Control's say in their daily activities and outside communications. By this point, oxygen levels were about the same as at 4,000 meters above sea level. At these extreme altitudes, sleep can be disturbed and one must pause for breath even after walking a short distance. The workload of removing biomass was relentless, so in the end, necessity prevailed. In January 1993, oxygen was pumped in, probably saving the experiment. In February, the science committee resigned because they were frustrated with the lack of progress and felt their recommendations weren't being heard. [133,139-142]

The 3,800 species in Biosphere 2 were chosen to be compatible and easy to handle within each biome, though their interactions weren't modeled in detail. Instead, SBV had faith that ecosystems would self-stabilize, perhaps after some extinctions, as they had in Folsome's flasks. During the first two years, though, all of the pollinators including honey bees and hummingbirds died, leaving many plants unable to produce seeds. In all, nineteen of the 25 species of vertibrates died. [141] Among the plants, the morning glory proved an implacable foe. Welcome at first because its fast growth would remove carbon from the air, this stowaway threatened to overrun several biomes including the cloud and rain forests. By the first summer, when the bionauts realized the severity of this threat, there were literally tons of it to be cleared. It took the remaining year of closure to eradicate it. [133]

Concentrations of nitrous oxide (N_2O, also known as "laughing gas") rose steadily throughout the first closure. Most of it probably came from denitrifying bacteria in Biosphere 2's rice paddies. Biosphere 2 actually had a mechanism for dealing with this and other nuisance gases: a device called a soil bed reactor, which would pump air through the soil in the agricultural biome. Experiments in the Test Module had shown that soil microbes digested the trace gases. [124] Unfortunately, such digestion would also increase their metabolism and hence respiration. This would add CO_2 to the already unacceptable levels present in the atmosphere, so the bionauts decided not to turn on the reactor. [135]

The first crew worked together for at least two hours a day in the agriculture biome, took turns cooking, and each had one or more areas of responsibility, such as maintaining one or more biomes. A bionaut had helped design the basement mechanical systems of pumps, wires and pipes. During closure, he was in charge of maintaining them; this took about 30 hours a week. Two others maintained a laboratory and a small medical center within the biosphere. "Mission control," located immediately outside, was in nearly constant contact via intercoms and the walkie-talkie radios that the bionauts each carried. Although thousands of sensors throughout the structure linked into the computer network, some had to be read in person and logged in manual notebooks. Windows throughout the structure (including under the ocean) let five hundred tourists a day see what the crew was doing at any given moment. There was no privacy, except in personal quarters. [133]

The Aftermath

The first crew emerged on September 26, 1993, two years and twenty minutes after closure. A second crew of five men and two women entered on March 6, 1994 for what was supposed to be a ten-month closure. The following month, Edward P. Bass, who had by then spent $200 million building the facility, fired its managers and called in federal marshals to eject them. Days later, two bionauts from the pro-Allen faction of the first closure, citing concerns for the second crew's "safety," broke into the facility, smashing windows and breaking airlock seals. They were promptly arrested. [129,139,140]

The second closure ended in September, 1994, four months early, due to oxygen loss and trace gas contamination problems similar to those encountered by the first crew.

Bass recruited Columbia University to manage the project in 1996, and in 1999 the University agreed to match his $50 million commitment over the next several years. The focus changed permanently from closed life support systems to ecology. The new science steering committee immediately recommended a one-year "baseline" study where the facility would not be altered at all. This would provide a picture of what the ecology would do on its own. Thereafter, the various biomes were monitored under intentionally altered gas concentrations, humidity and temperature. Unlike Earth, Biosphere 2 allowed researchers to watch where any given element (such as carbon) went in the system. [143]

For several years, Columbia operated Biosphere 2 as an extension campus where students could earn credit toward a Master's in public administration in earth science, policy and management. The University had hoped to fund it as a National facility, but the United States Department of Energy review panel gave it a low priority. So in December of 2003, Columbia withdrew from the facility and moved its academic programs back to New York. Bass put the Biosphere 2 facility up for sale, attracting several bids from real estate developers looking to build bedroom communities for Tucson and Phoenix.[129,144] As of this writing, its fate remains unknown.

The Lessons

Biosphere 2 set world records by a wide margin for terrarium size, degree of closure and human survival time inside a sealed environment. It managed this feat with private funds at a quarter of the cost of a single Shuttle launch—a relative bargain. But for building off-world environments, its main lessons are cautionary, to wit:

1. First things first. If you want to keep people alive, start with a closed agricultural module and see if that works. Don't begin by trying to do everything at once, such as keeping a desert alive right next to oceans and cloud forests.

2. Don't rush. It makes no sense to cut corners on the groundwork of a project designed to last a century or more. Construction on Biosphere 2 should not have begun before the Test Module's results were in.

3. Build a little, test a lot. Three closures ran in the Test Module for a total of 31 days. There should have been one or more two-year closures before ground was even broken for the full facility.

4. Use replicate modules. Space Biospheres Ventures could have built several more Test Modules of various sizes, some housing different biomes and at least one replicating the first module. After conducting baseline studies on the individual modules, movable transition modules perhaps could have linked them together in various combinations. This approach would have provided far deeper understanding at a tiny fraction of Biosphere 2's cost.

5. Test all materials using a wider range of environments than is typical of Earth. For example, the CO_2 absorbed into the concrete faster than expected in part because the atmosphere ended up with 10 times more CO_2 than expected.[145]

6. Replace vision with hypothesis. The biospherians assumed things would work "if we got the right size."[124] This became a controlling vision that, combined with their haste to make it succeed from day one, meant building the biggest facility their budget would allow. But how important is size compared with other variables? The Test Module's success should have raised this question—and led to a change in plan.

7. Be humble. SBV's leaders believed they could overcome human frailty through force of will. Although desire carried them far, they often commanded when they should have asked, fought when they should have cooperated, forced when they should have experimented, and concealed when they should have disclosed. Their arrogance led them into a downward spiral of dishonesty and hubris that cost not only the entire project, but the credibility of a promising new field.

To live sustainably beyond Earth would mean living in the image of Gaia: in closed, homeostatic ecologies that continuously regenerate air, water and food with minimal fuss. Biosphere experiments have shed some light on the process of closure, but only at great expense. Studies of global metabolism may illuminate the roles of climate, nutrients, biomes, niches and universal enzymes. However, as Biosphere 2 showed, Gaia's detailed arrangement of biomes may not provide a good template for closed human habitats. Using coral reefs and deserts to sustain people in a tiny ecology smacks of what Buckminster Fuller has called "special-case superstition."[146] Where are the invariants? What needs to be there and what doesn't? Which of the facts of the world also apply to a mini-world?

Perhaps we would benefit from an intermediate scale of experiment, somewhere between that of tiny closures and Gaia. If this line of inquiry proved accessible and rewarding enough to attract many researchers, it would stand a good chance of discovering new invariants much more detailed than Daisyworld.

Permaculture

As it happens, traditions that satisfy these requirements have abounded for thousands of years. A few have survived to historic times in such places as Australia, the New Guinae highlands, the Polynesian island of Tikopia, Ladakh

(Litte Tibet) and among native American tribes such as the Hopi. In each case, these cultures made clever use of indigenous resources to live in permanent partnership with their environment.

Unfortunately, there are few points of connection between the sustainable world of many indigenous peoples and our consumer society of glass, plastic, steel and advertising. To fill this gap and provide a bridge to living regeneratively in the modern world, two Australian naturalists invented permaculture.

When Bill Mollison and David Holmgren first coined the word in the 1970s, it was a contraction of "permanent agriculture." But among many of its thousands of practitioners world-wide, it has come to mean "permanent culture." Whether focused on humans or ecologies, permaculture is a design system that uses both indigenous and modern practices to live regeneratively within the local environment.

Unlike the industrial model of development, which uses specialization to build economies of scale, permaculture uses diversity to grow ecologies in place. It does this by observing and applying the strategies that successful ecosystems use to achieve their high cycling ratios. Invariably, this means increasing biodiversity.

The cascade of resources from one living organism to the next, so lengthy and varied in a rain forest, shrinks to just a handful of species in a desert. From this perspective, then, monocrop agriculture and even suburban lawns and homes qualify as intentional deserts: places where we eradicate diversity whenever it appears as insects or weeds. Permaculture rejects these widespread practices as failed design.

Successful permaculture teems with life at all scales, combining the bounty of a garden with the ease of maintenance of wild forage. It achieves this by redirecting the physical flows of resources and energy that were previously wasted or in conflict and sending them through an ever more complex cascade of uses by ever more living beings.

Permaculture's tools include built and landscape architecture. Interventions can be simple, such as imprinting conical pits in crusty desert soils to concentrate water, seeds and animal droppings that would otherwise blow away. Typically, though, permaculture combines a complex set of strategies, from solar architecture to small-scale waste water treatment systems to perennial polyculture. This is how Flowering Tree rapidly transformed its patch of desert into a lush oasis. Permaculture has been applied successfully to dwellings, landscapes, villages and bioregions from deserts to tropics to tundra.

Permaculture's Invariants

Because it is a relatively young practice, permaculture's invariants are still being discovered with some frequency. Practitioners often note and communicate these principles in the form of heuristics, or rules of thumb. Debates about the number or relative importance of the various heuristics are rare in Permaculture because solutions are local, and so is their relative priority. What matters in the end is not an invariant's place in a hierarchy of abstractions, but its efficacy in sustainable practice. One doesn't judge a permaculture design by the rigor of its precepts, but by its increased stability, biodiversity and yield as well as the decreased work it actually achieves on-site.

That said, the practice generally rests on two underlying themes: that we are not separate from nature, and that nature is fundamentally cooperative, not competitive. These form the ethical foundation of permaculture.

For example, permaculture embraces organic gardening, which focuses on the health of the soil, rather than waging war on it with pesticides and fertilizers. One of my gardening instructors, Lee Jones, has operated Stranger's Hill organic farm in Indiana for 30 years. She still occasionally loses a crop to insects. Yet rather than trying to eradicate them, she lets the crop go in order to attract beneficial organisms such as predatory wasps. While she's waiting for the natural balance to re-assert itself, she does not go out of business as a monocrop farmer might, because she grows a great diversity of crops every year. Not only does this strategy avoid the expense of buying chemical inputs, it rewards her with healthier, more productive soils over time.

Stories such as these abound in the permaculture world. Much of the strife we face in our struggle with nature comes from failing to recognize the cooperative patterns that nature already has invented. The main task of permaculture is to identify these patterns and apply them to designs for living.

Here are a few of permaculture's many heuristics that will prove especially useful to designing and building gaiomes:

Maximize the number of useful connections

With Daisyworld, Lovelock had shown how a greater diversity of species increases environmental stability. In the real world, permaculturists have noted that helpful diversity consists not just in the number of species but also in the number of beneficial connections and interactions among them.

Life negotiates its own connections, but it often falls to the permaculturist to make the necessary introductions. One of the simplest ways to increase connections is to increase the extent of boundaries between adjacent microclimates. For example, the edge of a pond provides a line of contact between aquatic and dry land environments. Numerous species interact there and it is possible to increase ecological connections by extending the line's length. Thus, a pond with a shallow, ragged, lumpy or wavy border will have a longer circumference and hence sustain a greater quantity and diversity of life than a deep, straight-edged pond with the same surface area. Similarly, a meandering river provides a longer path for the water, and hence greater and more diverse habitat areas than a straight-sided canal. For this reason, permaculturists often look to *increase the edge* in their designs.

Minimize work and pollution

Permaculture defines work as the mechanical or personal effort required to meet any need not provided by the local ecology. Similarly, pollution is any output that the ecology can't use.[147] While these would seem universal aspects of efficiency, industrial agriculture all but ignores them. For example, monocrops are vulnerable to weeds, pests, drought and disease. When combating these involves mechanized labor, irrigation, tillage and chemical pesticides, the machinery involved typically expends seven to ten calories of energy per calorie of food produced—a lot of extra work.[148]

Monocrop, tillage and pesticides disrupt the microbial community that supplies plants with nutrients and keeps the soil moist. Fertilizers and extra irrigation water must be brought in to compensate for these losses. Irrigation can deplete ground water reserves; fertilizers tend to leach excess nitrates into the remaining water supply, rendering it polluted and unsafe to use.

A permaculture approach to food begins by observing that forests and other natural ecosystems manage to feed thousands of animal species for millennia without work or pollution. Forest gardens are structured to work as much as possible like a forest, but with as many edible plants as possible. Once established, the main labor involved is foraging for ripe food, which reverses the food/labor caloric ratios of mechanized agriculture. The Polynesian islanders of Tikopia, for example, have used food forestry (among other practices such as fishing) to sustain themselves for thousands of years.

Even in Ladakh, an extreme high-altitude desert where forests cannot grow, the traditional culture thrived for a thousand years with minimal work and no pollution. In their four-month summer, they grew enough grain for a long winter of celebration and leisure. How? By irrigating carefully with glacial melt-water and replenishing the soil with a rich compost of wastes from their kitchens, hearths and latrines. Skill and attention were more important to their survival than labor: no one would drink from a stream designated for washing, nor wash in a stream designated for drinking water. When clothing became too ragged, it was used to line the irrigation ditches. There was literally no concept of trash in their society. [149]

Stack functions

Few things in nature have just one ecological function—in sharp contrast to engineering and science, which seek to isolate variables. Permaculture follows nature's lead by stacking functions. Every aspect of a design from the land contours to buildings to individual species should be situated to provide multiple benefits.

Ladakhis have long used the cold- and altitude-adapted *dzo*, a semi-domesticated yak-cow hybrid, for milk, labor, transportation and hair (for making blankets and shoes). Although it provided only a tenth as much milk as the jersey cow, its other yields added up to much greater value than a single-purpose milk cow could provide. The dzo also grazed in high pastures, requiring no special feed crop or shelter. [149]

Satisfy needs through diverse means

This is the basic engineering principle of block redundancy. If the weather or pests wipe out your potatoes, it's useful, even vital, to have alternative sources of food. Traditional horticulture in many parts of the world relied on a variety of species so huge that it can scarcely be comprehended in our commodified society.

Among the 2,000 native food plants cultivated in Africa, Kenyan gardeners have long grown root crops that could weather years of droughts and pestilence. As plantation agriculture invades that country under the false banner of relative economic advantage, governments have pushed for replacing this diversity with just a few cash crops, such as coffee. Not only are these introduced varieties not adapted to the local climate (thus requiring more resources such as water, pesticides and fertilizer), but their lack of diversity and food value is putting the survival of formerly self-sufficient people in the hands of very distant markets. The introduced crops function mainly to provide new customers for industrial agriculture companies.[48]

Even subsistence agriculture can go wrong by failing to value diversity. In *The Botany of Desire*, Michael Pollan describes how a million people starved to death in Ireland because they were all growing clones of the same potato, each equally vulnerable to a single fungus.[150] By contrast, the Incan civilization, where the potato originated, grew over 3,000 varieties in patchwork fields with open-pollinated edges. The same fungus that ravaged Ireland would hardly have dented the Incan potato harvest.

Use guilds and succession

Permaculture gardens rely heavily on perennial rather than annual species. Not only does this minimize the work and pollution inherent in tillage, it allows gardeners to make maximum use of the natural synergies found in communities of interdependent plant species. These associations provide not only numerous animal habitats, but also food and other materials useful to humans. Permaculture calls these beneficial associations *guilds*.

Unlike the agricultural practice of crop rotation, plant guilds provide many yields in the same space. Even simple annual guilds such as the native American "three sisters" (corn, beans and squash) stack numerous functions: nitrogen fixation, aeration, pest control, moisture retention, habitat for beneficial insects and food diversity. In *Gaia's Garden*, Toby Hemenway describes a much more complex guild of twelve or more species (at least nine of which are edible) anchored by an apple tree.[115] The diversity of species can reduce the gardener's work, their abundance of herbs, leafy greens and fruits maturing at different times to provide food forage practically throughout the year. What guilds don't provide, though, is easy harvest, rendering them unsuitable to mechanized agriculture.

There is a natural succession to plant growth that appears in the form of the pioneer species (annuals commonly known as weeds) that invade monocrop agriculture. Left unchecked for several years, these prepare the soils for perennial shrubs. In a few decades the shrubs give way to forests, which generally shade out the pioneers except along their fringes. By planting guild members in the right sequence, Flowering Tree was able to harness and accelerate natural succession to improve soils, conserve moisture and provide food and shade. [115]

Analyze flows by zone and sector

Permaculture analyzes the energy sources and sinks on a property according to generally concentric zones. [151] When planting a garden, for example, it's prudent to place often-used varieties such as herbs up close to the house, where they're easy to reach. Permaculturists call this the first zone. Plantings that need tending several times a week might be located in the second or third zone, farther away. Cash crops, firewood, native plants and game might be found in the third and fourth zones. Unmanaged wild areas that house beneficial insects and regenerate air, water and wildlife comprise the fifth zone. In densely inhabited areas, the first few zones may inter-penetrate each other, with the outer zones surrounding the entire settlement in as ragged a border as possible (per our first principle: maximize edge).

The microclimate around a dwelling varies considerably depending on the amount of exposure it has to sunlight, wind, rain, fire, noise and other material and energy flows. Because each flow generally has a preferred direction, a parcel of land can be analyzed in terms of sectors named for seasonal winds, the direction of sunlight, etc. A permaculturist will then seek to utilize as many flows as possible to advantage, and deflect or absorb undesired flows (such as noise). Evergreens might be planted as wind breaks against the winter wind sector, leaving gaps in the summer wind sector for a cooling effect. Houses might be built with slanting windows in the Sun sector to warm themselves in winter sunlight, and overhangs or deciduous plantings to provide shade in summer. A flood sector can be terraced or re-graded to divert water to contour irrigation ditches or a pond, which can provide water storage for irrigation during later droughts.

Ladakhis traditionally used dried animal dung to heat their houses during their fiercely cold winters. As Ladakh began to modernize starting in the late 1970s, imported coal (a non-renewable resource) could be had—for a price. But Ladakh has 300 days of sunshine a year. Anthropologist Helena Norberg-Hodge, recognizing this unused flow of energy, organized a demonstration project to fit Ladakhi houses with sun-facing windows sealed over a thick brick thermal mass. This upgrade paid for itself in reduced heating bills in less than two years.[149]

Cuba: A Permaculture case study

Let's see how these permaculture principles can work together to avert crisis and establish a more sustainable relationship with nature.

Cuba, an island nation with eleven million people, once depended on income from cash crops such as sugar, which it sold above market price to the Soviet Union. The country was far from self-sufficient, importing most of its rice and other food staples, and buying its oil at below-market prices from the USSR. Its huge, state-run sugar plantations were ultra-modern, using more machinery, chemical fertilizers and pesticides per-capita than the USA.[152,153]

With the withdrawal of Soviet oil in 1991 and subsequent embargoes that cut 80% of its exports and imports, Cuba suddenly became a gigantic experiment in self-sufficiency. Energy-intensive farming came to an abrupt standstill, and food rationing began. Before the food situation stabilized, the average Cuban had lost 13 kilograms.

During this "special period," Cuba decentralized its farms, allowing small cooperatives to grow and sell produce in farmer's markets (maximizing connections and diversifying supply). Diets shifted away from energy-intensive meat and dairy and toward more fresh produce (minimizing work and pollution). Permaculturists arrived from as far as Australia to provide educational assistance, especially in rebuilding the soils that had been depleted of soil microbes and other natural capital by chemical agriculture. In Havana, people began growing food on rooftops and any vacant parcel of land (stacking functions). Soon, such Zone 1 gardening would provide half the vegetables consumed in the sprawling city of 2.2 million people.

Today, Cuba is self-sufficient in food, 80% of it organic. The nation uses 21 times less pesticide per-capita than the USA and yields are improving as its soils recover from past abuse.

Permaculture vs. Globalism

One aspect of Cuba's successful move toward self-sufficiency should interest those for whom outer space would seem to offer an escape from tyranny. While it has long been argued that the solutions to the world's many resource crises will require global control and conformity, Cuba found the exact opposite to be true. By decentralizing farms, hospitals, markets, businesses and universities and collocating them with housing, Cuba not only saved fuel, it enlisted locals in their own welfare—and that of the land—without any loss of health, culture or scholarship.

A healthy ecology keeps resources in constant circulation. This is easiest when those resources don't travel far, but instead cycle through quick, short loops in the immediate environment. Living ecologically therefore tends to reduce global commerce and hence the need for global regulation. It promotes not only clean living, but local color, local wealth and local decision-making. In the solar economy, diversity and locality rule.

Permaculture solutions depend on local circumstances. Designs developed for hot coastal deserts will differ markedly from those in the temperate continental grasslands. Gardening the permaculture way involves heavy use of native and endemic plant guilds already pre-adapted to the local bioregion. Building the permaculture way arranges local materials to best effect with the local climate. If you travel among permaculture villages and towns, you'll find an amazing variation in architecture and landscapes.

Global unity and conformity are, if anything, a threat to permaculture. As a practitioner, you begin by examining the material flows that affect your family, town and bioregion. Then you get together and work out a design that makes the best and highest possible use of those flows. By implementing and improving on that design, you automatically reduce, eliminate or reverse your environmental demands on the global scale.

The world's food and energy crises are not amenable to pharaohnic solutions because pharaohs, by definition, must project their influence far beyond Zone 1. Not only does such oversight cost enormous amounts of energy, it also leads to serious mistakes: coffee replaces indigenous crops in Kenya; the order comes to launch the Space Shuttle after freezing rains.

Permaculture is anti-pharaohnic. It's all about noticing and applying patterns at the local scale. Simply put, it is the art and practice of becoming peaceably native to your local ecology.

Applying the Lessons

Our right of passage into the cosmos depends, paradoxically, on becoming better rooted here on Earth. In a gaiome, as with Gaia, the web of life would run full circle, but in much tighter loops. To live in space, we must once again become ecological beings.

It is humbling and ironic that most of those who live sustainably today are indigenous peoples whom modern civilization has conquered or marginalized. We moderns have much to learn from them about life in a world that does not revolve around human desire. Unfortunately for us, their connection to the land is too profound to graft directly into our culture of displacement. New Guinae highland children who go away to school often return to find they don't know how to garden.[50] Similarly, Ladakhi children return from city schools unable to saddle a dzo or weave (and hence repair) their own shoes.[149] As our culture assimilates them, they lose their ability to negotiate a satisfying life with their native environment, becoming dependent instead on economies that span the globe.

Permaculture aims to reverse this trend by establishing closer connections with nature and our neighbors. Some are starting to refer to this process as *relocalization*. By combining a few of the hard-won empirical experiences of indigenous peoples with direct observation and the results of penetrating, reductive science, permaculture can weave tighter links between people and their immediate environment. In most cases, this begins with restoring soil fertility in areas that, like the farms of Cuba and Kenya, were all but sterilized by the extractive economy.

We have much to learn before small, modern communities can thrive without waste, but some trends in that direction have already appeared. Publishers have long-since switched to soy-based inks. The construction industry is starting to take an interest in natural, renewable building materials. Manufacturers are beginning to realize economic gains as they move toward zero-emissions production.[154] While today most items from clothes to cars are made in specialized, massive factories, we are also starting to see a new crop of small-scale, precision fabrication facilities or "fabs" that make computer-aided manufacturing possible at the desktop scale.[155,156]

The day may come when a village of a few hundred people can make, maintain and recycle its own housing, clothes, transport and communications infrastructure. Food, no longer a trade commodity, would come primarily from Zone 1 and 2 gardens a few steps from home. With the human footprint thus reduced, much farmland can revert back to natural prairie, bog and forest. Without dependence on material imports, information from songs to science to fab recipes could be shared freely and globally. As more of these ecovillages come into existence and permaculture reduces work and pollution, enormous amounts of time and energy would become available to devote to the arts and sciences. Only such a grounded yet exuberant culture would have the skill and fortitude to build gaiomes.

Still, ecological cycles never entirely close, not even on Earth. Island forests far from continental deserts, migration routes and volcanoes gradually become depleted in minerals, resulting in ever more fragile ecologies (such as on Easter Island).[50] Continents often fare better, relying on volcanism, salmon runs and sea birds to replenish nutrients such as calcium, which gradually leaches away in runoff even from healthy forests.

Because of the gaiome's smaller reservoirs, cycles that fail to close would degenerate into crisis much quicker than they do on Earth. Without Earth's plate tectonics, oceans and volcanoes, a gaiome would need some other way to recycle calcium. Lacking a stratosphere with intense ultraviolet light, it would need some other way to break down the nitrous oxide that electrical equipment can generate. Ultimately, it would need machinery and human intervention to break down accumulated toxins and recover all vital elements, because at least at first, all cycling ratios will fall short of infinity.

Why then should we even bother to use natural ecosystems to support us in space? First, because it's part of who we are. When we turn away from living nature on Earth, our lives spiral out of balance, demanding ever more work, extraction and disposal as we struggle to overcome our expanding alienation. In the emptiness of space, healthy ecologies will keep us whole. Second, natural ecology amplifies the effect of mechanical life-support in proportion to its cycling ratios, many of which can run into the thousands or tens of thousands. With ample soils, water and greenery, a gaiome's chemical reactors can thus be orders of magnitude smaller, simpler, cheaper and easier to repair, move and build in quantity. They would need less power and produce far less waste heat.

Natural ecosystems have evolved in a larger world than we can build, on time scales longer than we can use. Therefore, we can't expect them to close all the loops in our gaiomes perfectly. Not right away. Still, we would be wise to bend every effort to make life the centerpiece of life support, because it embodies a wide range of technologies that we won't be able to duplicate for a long time, if ever. Earth's ecologies have been under continuous development for billions of years. With research budgets tens of thousands of times the current biomass of the human species, the forces of evolution and environment have relentlessly applied selection pressures to every living organism. As a result, our world is filled with beings and ecosystems that have survived the gauntlet of billions of parallel steps of optimization for energy and material efficiency, self-repair, self-replication, adaptation and flexibility spanning an enormous range of changing environments.

Our love of mechanical efficiency runs deep, and nothing in our brittle monocrop world can outshine the robustness of forests, flowers, pollinators, fish, amphibians, birds, insects, reptiles and all the unseen microbes that support us. If only for the comfort of their company, those who venture beyond Earth should bring as many living beings with them as they can, in as natural and wild an ecology as possible. But where should they go?

Chapter 4

New Worlds, Found and Made

"The most common things are the most useful..."
—William Penn (1693)

Wherever people choose to settle beyond Earth, local conditions will play a major role in shaping the details of their lives. Each place would provide settlers with unique degrees of access to energy and minerals that they can convert to dwellings, sunlight, air, water, ecosystems, food and amenities. Their location, whether near or far from Earth or the Sun, whether rich or poor in materials, will play a large role in defining not only the means of their survival, but also the terms of their relationships with Earth and each other. So will the uniqueness or multiplicity of these places.

This chapter surveys the solar system's diverse environments from the viewpoint of would-be settlers. By examining the various schemes that have been proposed to exploit their resources, it becomes strikingly clear how each environment shapes human values.

The Moon

The Moon is both farther and nearer than one might think. When asked to draw the Earth and Moon to scale on a sheet of paper, most students depict them much closer together than they really are. An accurate sketch (Figure 4.1) leaves no room on the page for any detail on the Moon. The Moon is one-third the size of Earth, and thirty Earth diameters, or 380,000 km away.

Figure 4.1: The Earth and Moon. Scale: 1 mm = 5,000 km. The outer line about the Earth depicts the Shuttle's highest orbit (1,000 km).

Though it is far beyond the reach of the Shuttle, the Moon is still, by a wide margin, the nearest major object in the solar system. By radio, it's only 1.28 seconds away. By rocket, it's never more than three days away, and every day presents several launch opportunities.

To reach the Moon from LEO, you need a ΔV of 3 km/s. But that's a crash landing; slowing down enough to get there in one piece requires another 3 km/s. Now, you might think a return to Earth would take another 6 km/s, but it actually takes only 3—if you use Earth's atmosphere rather than rockets to decelerate upon returning to Earth (this is called *aerobraking*). So it takes about 9 km/s to get from LEO to the Moon and back, about the same ΔV as it took to get to LEO from the surface of Earth.

The Moon is covered with regolith, a ubiquitous fine-grained powder created by billions of years of meteor impacts. By itself, regolith makes great shielding material; all you have to do to form a shelter is scoop it up and tamp it into hollow plastic forms. As an ore, the regolith is rich in silicon, oxygen, aluminum and titanium. One can, with extreme expenditures of energy, use it to make glass, metals and gaseous oxygen.

Unfortunately, the Moon is dry. The regolith has essentially no hydrogen, nitrogen or carbon. The stuff of life—water, organics and air—is scarce on the Moon.

Another real-world problem on the Moon is getting power. With a rotation period equal to its orbital period, its days last 14 Earth days—and so do its nights. If you want to use solar power, you must either store a fortnight's worth of energy (an extremely formidable task), or somehow get your ecosystem to go dormant for two weeks out of the month. If, instead, you rely on nuclear power, you then must deal with long-term waste handling and non-renewable fuels, which compromise the design goal of sustainability.

But sometimes a real world, in all its complexity, offers a helping hand. For example, in the 1990s, the Clementine and Lunar Prospector spacecraft, along with ground-based radar, have provided some evidence for water ice near the

Moon's poles.[157] This was surprising because every lunar day, the lit side heats up to 250°C, which should boil away any surface water. But in the Moon's polar regions, sunlight arrives at a low angle, spreading its energy out over a wide area, just as it does in the arctic regions of Earth. As a result, the Moon's polar regions are perpetually cold: the floors of some craters, particularly on its South pole, may not have seen the Sun in billions of years. During this time, many comets have struck the Moon. Comets are mostly made of frozen water and CO_2, so when one hits the Moon, it vaporizes, giving the Moon a temporary atmosphere. The molecular speed of this atmosphere is enough to gradually escape the Moon when heated during the day, but perhaps some of it condenses in the cold polar regions. With multiple impacts, it might even gradually build up a sizeable ice deposit. Unfortunately, continued radar observations from the ground have found daylit areas that look as "icy" in radar as the perpetually darkened areas.[158] This has somewhat doused hopes for large ice deposits on the Moon, though the discussion remains far from closed.[159]

Even without ice, it would make sense to site a lunar base near the poles because, just as some crater floors may have been in perpetual shadow, some crater rims may be in nearly perpetual sunlight. By mounting solar panels above an exposed rim, a base could remain powered up and productive throughout the month, neatly avoiding the dark fortnight that plagues the rest of the Moon.

Life in the Lunar Caverns

The Moon does not have enough gravity to hold an atmosphere, so lunar bases would need to be enclosed and air-tight. Though a domed metropolis in a lunar crater makes a pretty picture, it's not a very likely scenario. Without the protection of Earth's magnetic fields and thick atmosphere, a Moon base would need two to three meters of shielding from cosmic and solar radiation. So it's more likely that Moon bases would be built underground, perhaps in lava tubes, or inside large vaults of bagged regolith.

In his radio broadcast from Lunar orbit on December 24, 1968, Apollo 8 Commander Frank Borman remarked that the Moon seemed a "vast, lonely, forbidding type existence... and it certainly would not appear to be a very inviting place to live or work."[8] The Moon is undeniably desolate and colorless. On the other hand, Earth would loom three times larger in its black, starry sky than the Moon does in ours. Given enough dome space, trees might grow to many times their normal Earth height—though admittedly no one knows how the entire web of life, from tiny soil microbes to large predators, would fare in the low-gravity environment. At one-sixth gee, even small children could dunk a basketball on the Moon. The Moon is a truly different world, perhaps even a viable one, just three days away.

The amount of prime real estate on the Moon is truly miniscule, concentrated, if anywhere, in the high and low ground near the poles, where constant sunlight and perhaps lunar ice may be found in close proximity. If a lunar ice deposit actually exists, its alternating layers of ice and ejecta would record the history of the Moon and its impacts, making it a unique scientific resource. Like the much larger continent of Antarctica, the lunar poles may end up under international control, hosting mainly scientific endeavors, but little or no commerce. Although polar water would be welcome aboard a habitat, mining it may never be legal.

If commerce does become legal in the polar regions, their limited extent may eventually invite conflict between rival companies and nations, restarting the cycle of strife that plagues our world.

A new potential source of energy, Helium-3, would allow fusion reactors to operate while giving off minimal radiation. While rare on Earth, Helium-3 is found in the solar wind, and over time gets deposited and concentrated in the Moon's loose regolith. A number of schemes have been proposed to strip mine large tracts of the Moon's surface, extracting Helium-3 for shipment to Earth. Here again, a resource that renews on cosmic rather than human time scales is supposed to solve the world's scarcity problems. I have my doubts.

The Planets

The solar system has two distinct regions: the bright dry Inner Zone with its rocky planets (Mercury, Venus, Earth and Mars), and the wet, cool Outer Zone, with its gas giants (Jupiter, Saturn, Uranus and Neptune) and small, often icy worlds (their moons, as well as dwarf planets such as Pluto, Sedna, Quaoar and Eris).

The distances between planets are huge. Figure 4.2 shows the orbital paths about the Sun of Mercury, Venus, Earth and Mars. In this drawing, the Moon's orbit around Earth would fit within the thickness of the line used to draw Earth's orbit. Figure 4.3 shows the orbits of Jupiter, Saturn, Uranus, Neptune and

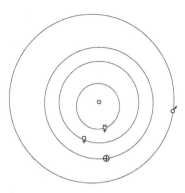

Figure 4.2: The orbits of the Inner Planets. Scale: 1 mm = 10,000,000 km.

Pluto. Note the change in scale: The outermost orbit (Mars) in Figure 4.2 is a third the diameter of the innermost orbit (Jupiter's) of Figure 4.3.

Mercury, the planet closest to the Sun, is very hard to reach, requiring a ΔV of 7.8 km/s each way. It's also probably as dry and depleted in organics as the Moon (though again, with radar evidence of an ice cap), so it may be safe from human interference for a long time to come.

Venus is almost as big as Earth. With a similar density, a much thicker atmosphere and no moon, it's harder to escape from than the Earth. Venus' atmosphere is toxic—nearly pure CO_2—with an instantly crushing pressure of 90 Earth atmospheres at the surface. At 0.7 AU from the Sun, it gets twice the sunlight as Earth. Worse, CO_2 is a greenhouse gas (it's opaque to infrared radiation), so the surface temperature is 737 K, hot enough to melt lead. Venus, too, will remain desolate for a long time to come.

The Outer Planets—Jupiter, Saturn, Uranus and Neptune—are much too far away for early development, as they require one-way trip times of five to fifteen years with near-term technology. Over the next few centuries, they may see the occasional scientist or space adventurer, but they're far too hard to reach and energy-poor to settle any time soon.

Pluto, together with its oversize moon Charon, is perhaps the most famous member of the Kuiper Belt, a swarm of icy dwarf planets that extends out to 50 AU. While about a thousand objects between 10 and 1,000 km across have been found in the Kuiper Belt to date,[160] they are too far away to concern us for a long time to come.

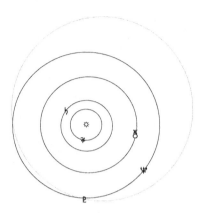

Figure 4.3: The orbits of the Outer Planets. Scale: 1 mm = 225,000,000 km.

Mars

Mars orbits the Sun at an average distance of 1.52 AU, which in practical terms places it over 1,000 times farther away than the Moon. Yet because Mars has an atmosphere, it's possible to land small spacecraft there using aerobraking, a parachute, and small rockets. In theory, it actually requires less propellant to reach than the Moon (though much more to return). In practice, though, landing people or large equipment on Mars would involve aeroshells (combined aerobrake/heat shields) much larger than would fit on the current stable of rockets.

Mars has a synodic period (i.e. mean time between best launch opportunities from Earth) of 27 months, and a trip to Mars takes six to nine months. Round trips take multiples of 500 days: once you land on Mars, you'll have about 60 days until your next opportunity to return to Earth. If you miss that flight, you're stuck on the Red Planet for another 25 months.

Yet for all the logistical hassle of getting there, Mars offers several distinct advantages over the Moon. Its land area is nearly the same as the dry land area of Earth, or three times as much as the Moon. It rotates once every 24 hours 37 minutes, perhaps close enough to an Earth day to make living there comfortable. Mars gets half the sunlight per square meter that we do on Earth, and its record high temperature at the equator in the summer is only 20 °C. Its atmosphere, a hundred times thinner than Earth's, does little to prevent warmth from escaping the planet directly into space. However, its mainly CO_2 composition is a great potential resource for Earth life, which needs both carbon and oxygen to survive. Earth-based telescopes have been able to detect water ice on the Martian ice caps for decades, and spacecraft surveys are slowly accumulating evidence for subsurface water in other locations. No one knows yet whether its surface gravity of 0.38 gees (twice as much as the Moon's) is enough to maintain muscle, bone and ecologies imported from Earth.

Mars, like most planets, is differentiated: early in its history, it melted, causing most of the heavier materials—mainly the metals—to sink to its center, forming a core. (The cores of the Moon and Mars have cooled and solidified, while Earth's is still molten.) The outermost portion of a planet, called the crust, is made of lighter, metal-poor minerals. This is why metals are scarce on Earth. They are likely to be scarce on Mars, and (because it lacks Earth's long-term aqueous and biological processes) distributed differently than they are on Earth. Prospecting on Mars will be challenging. Mars has enough atmosphere to have weather, including powerful storms that deposit incredibly fine-grained, iron-rich dust everywhere. But the air is toxic to humans and much too thin to breathe. Its pressure is so low everywhere except in its deepest valleys that if you opened a bottle of water there, it would boil until the remainder froze.

Martian *areology* (from Ares, the Greek word for Mars) is so different from Earth's geology that it could take a long time to locate any economic mineral deposits it may have. To give a sense of the differences, it shows no evidence for plate tectonics, yet its tallest mountain (a shield volcano) is five times higher than Everest. Despite the dust storms, millions of impact craters dot its surface, preserving a record of impacts from millions of years in the past. By contrast, Earth's weather has obscured all but just a couple hundred impact craters. Mars has been dryer than Earth's driest deserts for billions of years.

Settling Mars would be entirely different from settling the Moon. The 8- to 40-minute communications delay makes two-way voice conversations with Earth more like a mail correspondence than a telephone call. A round-trip to the Moon takes only a week, while a journey to Mars requires a commitment of years. Because of its distance, Mars gets half as much sunlight as the Moon, but distributed in something much closer to the daily ration we're used to on Earth. The Red Planet is chemically richer than the Moon, but its enormous landscape was shaped by very different processes than Earth's. Mars is a big world that will probably take a long time to get to know.

The Martian atmosphere is far too thin to provide any protection from cosmic and solar radiation. So, like Lunar habitats, Mars bases would require shielding in addition to air-tight enclosure. Despite its much greater abundance of water than the Moon, Mars is still much dryer than Earth, and water would need very careful management. Plants can liberate the oxygen from the CO_2 in the atmosphere (when pressurized), but they still would need nitrogen, a source of which has not yet been identified on Mars.

The discovery of vital mineral and organic deposits would surely fan the hopes of prospective settlers. However, if these deposits turn out to be localized or rare, they will, like the Moon's ice deposits, become focal points of contention. The struggle to control these strategic assets could dominate the future of Mars for a long time to come.

Terraforming

The dream of living and breathing in the open air of Mars was a staple for over a century of science fiction. Authors from H.G. Wells to Robert Heinlein depicted Mars as a living world, albeit an older one with perhaps a more fragile ecology than ours.

Such dreams collided hard with the Space Age. When Mariner 4 flew by Mars in 1965, it managed to snap 21 pictures of the Red Planet. While they were a thrilling technical achievement, the pictures also must have left a sinking feeling in the stomachs of those scientists who grew up on a steady diet of John Carter, *Warlord of Mars*. The pictures revealed an old, dry, cratered surface that resembled the Moon far more than it did the Earth. Never mind the hints of frosts, clouds and weather patterns. While Mars was a much more colorful and dynamic world than the Moon, its stark terrain paled in comparison to the "grand oasis" of Earth.

Still, it was not long before serious scientists were again discussing a verdant Mars suitable for human life. This time their vision lay not in fictions past or present, but in future engineering projects. Astronomer Carl Sagan first brought up the subject of "Planetary Engineering on Mars" in a brief technical note in 1973.[161] Noting that the water of Mars seems locked up in the frozen polar caps, Sagan suggested melting them by importing "some 1 to 10 tons of low albedo material" to cover the caps. Warmed by the Sun, this dark material would sublimate the caps, adding enough water and carbon dioxide vapor to the atmosphere to raise its pressure. Eventually, the atmospheric pressure would reach the point where liquid water could remain on the surface without boiling away. This would achieve the goal of "rapidly transforming Mars to much more earth-like conditions." Sagan next suggested introducing dark plants which grow on polar snows.

By 1979, scientists were holding conferences on *Terraforming*: the process of making planets more Earth-like.

To accommodate Earth life, the atmosphere of Mars would need to be thicker, warmer and wetter. To achieve this, physicist Freeman Dyson, borrowing from a 1955 Isaac Asimov story,[162] suggested sending autonomous, self-reproducing robots to mine the ice of Saturn's moon Enceladus. The little automatons would then travel to Mars and crash their ice cargoes like meteors on the polar caps,[163] a strategy that I am happy to report also worked well in the 1990 Maxis game *SimEarth*. (Dyson did hasten to point out that his Mars scenario was a mere thought experiment to illustrate the uses of self-reproducing machines).

Soon enough, even James Lovelock, inventor of the Gaia theory, became engaged in terraforming thought experiments of his own, though he preferred to call it *ecopoesis* ("making a home"), a word that sounded less industrial to him than terraforming.[112]

The Minor Planets

The solar system is filled with debris left over from the formation of the planets 4.55 billion years ago. Only a tiny fraction of these objects are large enough or ever get close enough to be seen. Most of the known asteroids (also called minor planets) orbit somewhere between Mars and Jupiter, while most comets come from a zone far beyond the visible planets. Asteroids are generally rocky, while comets are icy. Objects smaller than about 50 meters are rarely large enough to be seen from Earth except when their impact is imminent. Such objects are called *meteoroids*.

When a meteoroid or the occasional asteroid enters Earth's atmosphere, it's typically moving anywhere between 40 and 70 km/s, or one to two hundred times faster than the speed of sound. The air molecules that it slams into on the way down form a supersonic compression wave that glows white-hot, often melting its surface. The light from this aerodynamic heating is called a *meteor*. Most meteors that you see at night are dust grains or pebbles that originated in comets. In the rare case where a meteoroid makes it all the way to the ground without boiling, ablating (flaking) away or crumbling under the sudden deceleration, the rock that survives is called a *meteorite*. Most of these came from asteroids. Through luck and organized searches, people throughout the world find several hundred meteorites each year.

Meteorites span a wide range of compositions. Some 94.1% of them, the *stones*, look like oddly weathered Earth rocks. Another 5.2%, the *irons*, look nothing like Earth rocks: they're made entirely out of nickel-iron and traces of other metals. The remaining 0.7%, the *stony-irons* include both rocky and metallic material.[11]

Most (90%) of the stones have never experienced melting of the sort that planets such as Earth did. Because they contain tiny *chondrules* (glassy or crystalline beads), we call them *chondrites*.[11]

Asteroids

Studying meteorites in a chemical laboratory would seem a straight-forward way to figure out what the asteroids are made of. However, the relationship is less obvious than may seem. Most of what we know about asteroids is based on telescopic observations, where they appear as very faint points of reflected sunlight. Their spectra or even relative brightness in different color filters tends to sort them into a handful of different classes. But the relationship between any given asteroid class and any type of meteorite has been punctuated with controversy for many decades. Through a telescope, the reflected colors of many of an asteroid's surface minerals can blend together. Space weathering, caused by micrometeorite impacts, solar charged particles and cosmic rays, can darken the exposed rock. Regolith from impacts can hide the rock from view. Each of these processes can flatten the spectra, hiding features that would be obvious in a surface sample or a meteorite found on Earth.

Early spectral classifications were based on what observers thought the asteroids were made of. M types were metallic, S were silicaceous (stony), and C were carbonaceous. Today, at least nineteen asteroid spectral classes exist, and scholarly tables linking them with meteorite types are festooned with question marks. Asteroids have a wide range of compositions and mineralogy, which makes sense considering their wide range of sizes and orbits. Most of the asteroids never got big enough to differentiate like Earth and Mars. These would include the C-types, some of which may have remained far enough from the Sun to avoid having their volatiles baked out. Other asteroids may be frag-

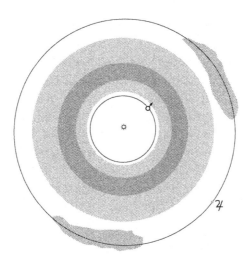

Figure 4.4: The asteroid belt between the orbits of Mars and Jupiter, with the Main Belt highlighted. The Trojan asteroids lead and trail Jupiter (shown not to scale) in its orbit. Scale: 1 mm = 20,000,000 km.

ments from larger bodies that did differentiate to some extent. These would include the M and S-types.

At first glance, the M-types would seem the most astounding and economically useful among the asteroids: tiny worlds made almost entirely of high-quality stainless steel. Recall from page 31 that the main belt holds thousands of M-types each with more tonnes of steel than humans have ever used. Unfortunately for the mining industry, though, the transportation costs are still forbiddingly high.

The majority of known asteroids orbit the Sun between Mars and Jupiter. Figure 4.4 depicts these *Main Belt* asteroids as a shaded ring. Another class of objects leads and follows Jupiter by about 60° in its orbit; these are its *Trojan* asteroids. Fewer Trojans have been found than Main Belt asteroids, but that's to be expected: the Trojans are nearly twice as far from the Sun and, under ideal observing conditions, roughly twice as far from Earth. Thus the average trojan appears some 16 times fainter than it would in the Main Belt. Factor in their generally redder, darker coloring, and the number actually detected to date hints at a population rivaling that of the Main Belt.

Although it was once popular to think of the asteroids as the remnants of an exploded planet, their mineralogy suggests that they formed in place as aggregations of dust, ice, pebbles and boulders in the early solar system. The total mass of the main belt asteroids is much less than the Earth. The largest of them, 1 Ceres (the "1" signifies its order of discovery), is only 1,000 kilometers across. Earth outweighs Ceres by a factor of two thousand. But most asteroids are small: there are 40,000 main-belt asteroids larger than a kilometer across,[164] and uncounted billions of asteroids smaller than that.

Main-belt asteroids tend to be found in families, but that doesn't mean you will find them close to one another. Two members of the same family can be on opposite sides of the Sun. What makes them a family is that they are at the same *average* distance from the Sun.

Life On or In an Asteroid

Over the years, numerous authors have suggested building habitats on or inside asteroids. Tsiolkovsky's stories from as early as 1895 abound with them.[60] Isaac Asimov's 1956 story *Strike Breaker* describes how an asteroid 150 kilometers in diameter, if hollowed out into levels 15 meters high, would have as much land area spread out over its 10,000 levels as all the continents of Earth.[165] Astronomers have found dozens of asteroids larger than this, so asteroids would seem to hold the potential to support, at the very least, several dozens of Earths worth of new land.

How would settlers alter the asteroids for their purposes? Some asteroid habitats might evolve from early mining efforts that produced tunnels, which then could be sealed and furnished. Other asteroids might be used entirely for habitats from the beginning. In their 1964 book *Islands in Space: The Challenge of the Planetoids*, Dandridge Cole and Donald Cox suggested the scheme to melt and then inflate metallic asteroids that landed Jeff Bezos in hot water at the start of this book.[166] Other authors have suggested enclosing parts of asteroids with domes or eventually bubbles that enclose them in their entirety.

Whatever scheme the authors employ, the allure is always the same: to obtain far more room for life to grow than is available on Earth. How much room is there in the asteroid belt? Obviously, that depends on the total number of asteroids.

Over the past 140 years, more than 600,000 asteroids have been discovered, and several thousand more are found each month.[167] Yet these are by no means the majority of objects out there; most are faint and remain undiscovered.

An asteroid's brightness depends on three things: its size, distance and albedo. Albedo measures the reflectiveness of a surface. A white surface (such as fresh snow) would have a very high albedo, while a dark surface (such as coal) would have a low albedo.

A given asteroid's size and albedo generally don't change, but because both it and Earth are moving, its distance can vary wildly. An asteroid that is easy to find right now might not be visible in two months.

Perfect sky surveys don't exist: telescopes generally look at only a tiny fraction of the sky at one time. The dimmer the object in question, the longer it takes for its photons to build up to a detectable signal in a telescope. From these facts, it is possible to estimate what fraction of objects of a given albedo and type of orbit a telescope should see over the course of a survey. If your survey is 1% complete for a particular type of object, then for every one you detect, you can assume there are 99 more out there.

Using this kind of reasoning (albeit with the benefit of over a century of surveys and much greater mathematical rigor), it has been estimated that the main belt includes around 8,000 asteroids larger than 10 km in diameter and 800,000 asteroids larger than 1 km in diameter. The total mass is something like 4×10^{18} tonnes, or about a thousandth as much as Earth.[14]

An object's surface gravity scales directly with its density and radius. With diameters typically hundreds to thousands of times less than Earth, and densities typically within a factor of two of Earth's, asteroids would provide scant gravity for their inhabitants. On the outside of a ten-kilometer S-type asteroid, for example, you would have to wait several minutes for an object to drop from your hand to the ground. With a single jump, you could easily soar hundreds of meters.

Converting an asteroid into a three-dimensional city-state whose citizens can soar through endless tunnels, vaults, domed craters or open spaces does have a certain storybook appeal. However, it doesn't seem likely that a natural ecology that evolved for billions of years in a full gee would do well there. Certainly humans would not: decades of astronaut and cosmonaut adventures have demonstrated how bones and muscles quickly degenerate in microgravity. How many of the thousands of species we depend on for life support would fare any better in milligravity? No one knows.

The deeper you go inside an asteroid, the less gravity you'll feel. Inside

Cole and Cox's hollowed-out asteroid, the only gravity would come from the air that was pumped in during construction, and it would pull you very slowly toward the center—away from its inner surface, the only available land. Cole and Cox solved this by rotating the asteroid habitat to produce pseudogravity along its inner walls (we'll talk more about this technique next chapter). But spinning up an asteroid takes a lot of work. Unless it's a pure, unfractured M-type, it would need significant structural reinforcements to hold together.

The asteroids have enormous potential for settlers, but they are farther away than Mars and proportionally harder to reach. At the distance of the Main Belt (2 to 4 AU), settlers would receive anywhere between four and sixteen times less sunlight than they would near Earth (compared to half at Mars). They also would need to expend enormous amounts of energy to render an asteroid livable for Earth ecologies. Because of these many technical challenges, settlers will not soon be scrambling to stake their claims on the main belt asteroids.

Comets

Several new comets appear in telescopic sky surveys every year. These loose "snowballs" of frozen volatiles with a thin admixture of dust are typically a kilometer or so in diameter. Many of them have extremely elongated orbits with aphelia of 10,000–50,000 AU (up to 1/5 the distance to the next nearest star). At this distance, it takes them millions of years to go around the Sun. That means for every one we see, there must be at least several million others "out there." Astronomer Jan Oort pointed this out in 1950. Since new comets come from every direction in the sky, Oort concluded that they came from a spherical cloud surrounding the Sun. Soon it became clear that reaching the 5 AU extent of the inner solar system from the *Oort cloud* was much less likely than scoring a hole in one in golf, especially since far from the Sun, the gravity of passing stars stirs up the comet orbits like a fierce, gusty wind. Detailed statistical studies suggest that for every comet orbit that reaches the inner solar system, there must be millions that don't. Thus the Oort cloud should contain *trillions* of comets.[168]

Comets are the most primitive debris left over from the formation of the solar system. The Outer Planets probably accreted in place from massive belts of comets that formed shortly after the Sun was born. The few comets that survived did so by being ejected via the Outer Planets' powerful gravitational slingshot. Many went into the inner solar system and boiled away. Most of the rest presumably ended up in the Oort cloud. A few ended up in the Kuiper Belt, now thought to be the source of the short-period comets.[160]

When comets first enter the inner solar system, the gravity of Jupiter and other planets can tweak their orbits, capturing them permanently. If their new perihelia are less than one AU, they typically last only a couple hundred orbits before they boil off all their surface ice and go extinct. What's left is most likely one or more icy nuclei covered in thick mantles of very black dust. Like C-type asteroids, these may have all the raw elements needed for life.

Comet Trees

The organic richness of comets was not lost on visionaries such as physicist Freeman Dyson. In his book *Disturbing the Universe*, Dyson noted that Earth, at the boundary between the solar system's bright-dry Inner Zone and its wet-cool Outer Zone, was rich enough in both water and sunlight to give life a good start.[163] In his view, life needs water more than sunlight, so the Outer Zone, with its trillions of comets, holds enormous promise in the long run. The problem is getting life established in the light-starved wilds.

Because he was gazing very far into the future, Dyson avoided specifics as much as possible. He imagined gigantic trees growing out of distant comets, with branches extending for hundreds of kilometers into open space. Huge, silvery leaves, space-hardened and bred for autotrophism, would concentrate sunlight enough for photosynthesis. Genetically-engineered roots would form insulated greenhouses that provide shelter, air and nutrition to humans and other symbiotes.

Every living thing in this new ecology would need heavy modification to survive. Human muscles and bones, unaccustomed to milligravity, would waste away on their own. So, presumably, would many among the hosts of other species from soil microbes to decomposers. We can't expect engineering alone to perfect their genetic revisions in a laboratory: life makes its own connections, and there are too many variables to control ahead of time. But centuries from now, a truly advanced engineering effort may at least give space-faring biomes a sort of limping start. Then, under selection pressures as mobile as their tiny orbits, the trees and those who dwell among them would begin to evolve.

Dyson is an expansive philosopher. It is no coincidence that the Outer Zone, so conducive to life, goes on forever. If engineered to reproduce, these trees would cast their seed among the comets, some of which would eventually leave the solar system by way of gravitational encounters with Jupiter or Saturn. Properly aimed by gentle nudges before each swing-by, these interstellar comets would encounter other solar systems within tens of thousands of years. By then, succeeding generations of tree ecologies would have spread into the Oort cloud and among passing interstellar comets, evolving to thrive in new niches as far-flung as the spaces around red dwarfs or blue supergiants. Once begun, there would be no stopping the spread of life far beyond the solar system and into the galaxy.

But all this is far beyond us for now. As the previous chapter showed, we don't yet know how to keep a closed natural ecology running here on Earth, let alone in space. The endless Outer Zone is an environment even newer and stranger and more varied than land once was, when life climbed onto it 500 million years ago. Before anyone can dwell there, we have a lot to learn closer to home.

NEOs

Thousands of tiny planetoids have found their way into orbits that cross Earth's: these are the Near Earth Objects (NEOs). Things happen fast in the inner solar system. If the NEO is icy, the Sun's warmth will boil out its volatiles in a few hundred orbits. Even if it's rocky, it will be ejected from the solar system or else destroyed through collisions with the Sun or a planet within a few million years.[169] The inner solar system has been around a lot longer than that, so where are the NEOs coming from?

There are places in the Main Belt where an asteroid's orbit would bring it close to Jupiter at about the same place every few orbits. When this happens, Jupiter's gravity deforms the asteroid's orbit relatively quickly, flinging it elsewhere in the solar system. Over time, this process clears gaps in the belt. Asteroids that wander into one of these *Kirkwood gaps* can be hurled outward or inward. Some pass close to Earth.

This is both bad and good news. On the one hand, these Near-Earth Asteroids (NEAs) are a perennial threat because every now and then one hits Earth. The most notorious example was the impact, 65 million years ago, that wiped out the dinosaurs. Smaller impacts are more frequent: a 1 km asteroid strikes Earth every 150,000 years, causing more damage than the combined nuclear arsenals of the world.[164] Smaller asteroid impacts happen even more frequently.

In 1908, for example, something exploded over Tunguska, Siberia with the force of a 10 megaton nuclear bomb. Meteors large enough to produce a blast comparable to a nuclear bomb probably occur several times a century, and a similar, though much smaller event may have occurred over Kazakhstan in the 1940s.[170]

The good news, of course, is that their proximity makes some NEAs easy to reach. Some years ago, I asked Professor John S. Lewis of the University of Arizona's Lunar and Planetary Laboratory about the resource potential of NEAs. Here's his response:

"A typical meteorite that falls on Earth has about the same composition as planet Earth put through a giant blender—the core homogenized with the mantle and the crust. So anything on Earth that we think of as core material, principally ferrous metals, would be vastly more abundant and more accessible on these asteroids. Many of these asteroids are carbonaceous, even some of the near-Earth asteroids. And the carbonaceous ones are just simply dripping with water and carbon compounds. The amount of volatiles in them that is easily extractable ranges upward of 40% by weight."[171]

The carbonaceous or C-type asteroids that Professor Lewis was talking about are very dark with carbon—darker even than coal. Therefore, they are much harder to find in a telescope than the bright S-types that dominate the catalogs of NEAs and main-belt asteroids alike. Extrapolating the size distribution of the C-types into the hard-to-detect range suggests they may comprise as much as 75% of the main-belt asteroids, mostly in families 2.7 AU or farther from the Sun.[11]

C-type asteroids were originally named for the carbonaceous chondrite meteorites that show similar spectral features. These are very primitive objects that probably have never been heated above $200\,K$, so they still retain lots of water and other volatiles. Unlike the Moon and Mars, each C-type asteroid contains, in a small, loosely packed mix, all of the elements needed for life, as well as enough metals and silicates to build a habitat.

Table 4.1 gives a sense of the differences between the various NEOs and the Moon by comparing the abundance of key elements in each as a percent of total mass. The relative population of Earth-approaching comets comes from published NEO discovery statistics for the comets.[172] Unfortunately, very few

Object type:	Asteroids			Comets	Moon
	C	S	M		
% of total population:	71	20	8	1	N/A
Major Life Elements (mass %)					
Hydrogen	2			6.4	
Carbon	3.5			19	
Nitrogen	0.08			2.1	
Oxygen	46	38		51	42
Major Tech Elements (mass %)					
Magnesium	9.7	15		3.8	5
Aluminum	0.9	1.2		0.3	4.7
Silicon	11	19		8.1	18
Iron	18	22	70	4.6	13

Table 4.1: Elemental abundances as a percentage of total mass for Near-Earth Objects. Quantities significantly below 0.1% are left blank.

spectroscopic studies have been made of NEOs, so the table uses Lewis's main-belt estimate for the relative population of C-, S- and M-types.[11] The asteroid compositions are estimates based on published laboratory studies of C1-chondrites, L-chondrites and Ataxite (iron) meteorites; the comet data come from observations of periodic comet Halley; the Lunar data come from Apollo mission samples of Titanium-rich mare basalts.[173]

Clearly, the numbers quoted here are at best indicative and certainly not definitive. Not only are they based on incomplete data and estimates, but compositions vary significantly (by 20–40%) among meteorites and from point to point on Lunar, asteroidal and cometary surfaces. Nevertheless, the gross trends are clear. The Moon is well-supplied for construction but not for life. S-type asteroids lack only nitrogen and hydrogen. Comets stand out for their wealth of all the major elements needed for life and technology, but are not commonly found near Earth. The most common objects, the C-type asteroids, have everything too, though perhaps less nitrogen than comets.[174]

About 25% of all NEOs require less propellant to reach than the surface of the Moon.[164] Because their surface gravity is negligible, nearly all of them are much easier to return from than the Moon, which takes a 3 km/s toll during ascent. There are probably about 2,000 NEOs larger than 1 km across, plus an additional several hundred thousand larger than 100 m.

While only about 1% of the NEOs have been identified as comets,[172] Lewis and others have long suspected that some of the C-types could be comet nuclei that have expended their surface volatiles.[164] If this notion proves out, these asteroids would have the ideal mix for world-building: carbonaceous, metal-enriched crusts surrounding veins rich in nitrogen and other life elements.

The Earth-approaching periodic comets tend to have eccentric orbits with aphelia reaching nearly out to Jupiter. They are poor candidates for early habitation because at aphelion (most distant point in the orbit), they receive some 25 times less sunlight than we do near Earth.

The near-Earth asteroids fall into three families: the *Atens*, with average orbital radii less than 1 AU; the *Apollos*, with orbits larger than 1 AU yet eccentric enough to cross Earth's orbit; and the *Amors*, which orbit between Earth and Mars, some perhaps crossing Mars's orbit.

Of the estimated 2,000 NEOs larger than one kilometer, up to 1,400 could be C-type asteroids. If half of these prove chemically suitable for life and machinery, then statistically speaking about 180 of them should take less rocket propellant to reach than the Moon.

NEOs do have two serious drawbacks for would-be settlers: long trip times and a long and irregular wait between launch windows. NEOs are in constant motion, of course, and they spend most of their time far from Earth. Were their orbits circular and not inclined much relative to Earth, their launch and departure windows would follow their synodic periods (the mean time between conjunctions with Earth). But in fact most NEOs travel in very eccentric and often inclined orbits, making their close approaches less regular. For example, asteroid 4660 Nereus has a synodic period of 2.22 years, yet it won't get close to Earth until 2021, when it will pass within 0.026 AU—about ten times as far as the Moon. Its next close pass (0.11 AU) won't come until 2031. Yet the asteroid currently known as 1996 FG3, which has a synodic period of 13.2 years, will have close approaches in 2009, 2011, 2022, 2024, 2035 and 2039.[175] While low-ΔV launch windows don't coincide with closest approach, they are just as erratic.

A very small number of asteroids, the so-called *Arjunas*, have "Earth-hugging" orbits with low inclination and eccentricity. These asteroids require very low ΔV to visit—at the cost of a very long synodic period. However, the windows of acceptable ΔV to the Arjunas remain open for years. During this time, spacecraft from Earth could make several nearly straight-line visits before the asteroid slowly drifts out of range, not to return for several decades.

Overall, though, the NEOs pose fewer challenges for settlement than the main belt asteroids: they are much easier to reach and they get far more sunlight. However, settlers would still need to deal with their milligravity. While it may be possible to reinforce and spin them up, another approach would be to use the NEOs as raw materials for free-floating habitats.

Built Worlds

In 1969, Gerard K. O'Neill asked his Princeton students whether Earth really was the best place for civilization in the long run. When they considered the world's many natural catastrophes such as floods, droughts, asteroid impacts and disease, along with its limited fossil fuel reserves, they began to question planetary life's long-term viability in the long run. Our survival thus far seemed a lucky accident. Should our future also be left to chance?

The class considered other planets, but quickly realized that the combined area of the Moon and Mars would not support more than one or two additional doublings of the human population—a limitation that their own generation might face.

In the view of O'Neill and his students, the problem with planets was their innate inefficiency. If Earth were an apple, they reasoned, its entire biosphere, from the outermost breathable wisps of atmosphere to the deepest inhabited oceans and soils, would occupy a volume thinner than its skin. The remaining 99.9% of the world's mass would appear to contribute nothing to life except the gravitation that holds in the atmosphere.[107] Indeed as early as 1907, Tsiolkovsky commented that if Earth's entire volume of material were divided equally among every living human, it would comprise 1.3 billion lumps, each of which, if spherical, would measure 11.75 kilometers in diameter.[5] If we could repeat the experiment today, with 6.5 billion of us, your individual share would measure only 3.5 kilometers across. But that's a lot to work with. Since it takes only a few centimeters of metal or woven glass to contain an atmosphere, and a few meters of soil to screen out biologically damaging cosmic rays, it should be possible to stretch out the material of each person's chunk into new, hollow biospheres. "Your" biosphere alone would thus be large enough for biomes that support tens of thousands of people indefinitely. To generalize the concept, built worlds should support orders of magnitude more life, tonne for tonne, than natural worlds.

O'Neill and his students soon became the space-age champions of Tsiolkovsky's vision. Unlike von Braun's antiseptic paramilitary space stations, O'Neill colonies would be artificial worlds designed to correct all the flaws of Earth, or indeed of life on any world, with all its conflict and scarcity. The colonies would involve engineering projects of unprecedented scale: spherical and cylindrical constructions several kilometers across, each housing thousands and eventually millions of people.

In O'Neill's colonies, everything would be controlled. The weather, climate, agriculture, recycling, etc. would all be handled through meticulous large-scale engineering. The colony's architecture would ensure that earthquakes, volcanoes, floods and famines never threatened the populace. Centripetal acceleration would provide the feeling of Earth-normal gravity in most areas, though people would no doubt invent low-gravity activities near the hub.

The scale of these colonies ruled out building them from materials hauled up from Earth. The smallest O'Neill colony, the Stanford Torus, would have massed ten million tonnes. Even at the fabled $260 per kilogram to orbit that NASA promised for the Shuttle in 1971,[176] simply launching the raw materials into orbit would cost $2.6 trillion in the currency of the day (equivalent to $12.6 trillion in 2005). So O'Neill and his collaborators looked to the Moon as a potential source of supply.

As mentioned on page 114, the lunar regolith is rich in titanium, silicon, aluminum and oxygen, though it is utterly depleted in hydrogen and nitrogen. One could, with extreme expenditures of energy, get glass, metals and oxygen from the Moon. Lunar regolith by itself makes a great shielding material, and all you have to do is scoop it up. Because shielding accounts for the largest portion of a space habitat's mass, O'Neill and his collaborators looked for a way to get it off the Moon efficiently and cheaply.

They found one. The idea was to scoop and bag the regolith, then put the bags in buckets that travel along a track perhaps several kilometers long. Magnets would accelerate the buckets to escape speed, unhindered by atmospheric friction. They called this arrangement a *mass driver*. Theoretically, given enough electrical power, a mass driver could fling objects all the way across the solar system.

Of course, there must be some sort of "catcher" on the receiving end.

O'Neill envisioned a mass driver in reverse. In such a system, both the "pitcher" and the "catcher" must be fairly complex. The pitcher must account for the ever-shifting position of Earth and the Moon and the location of the catcher in its orbit. Once the catcher is full, it needs to navigate its way to the construction site, drop its cargo and refuel. This sort of autonomous navigation and rendezvous technology is only now under serious development.

While a mass driver system would involve the high cost of putting heavy equipment on the Moon, it could also, over time, return many times its mass in shielding material to Earth orbit. The ratio of the mass returned to the mass of the lunar equipment, the catcher, and all fuels and crews involved is called the mass payback ratio. The higher the mass payback, the lower the overall cost. For example, a mass driver operating for many years might achieve a mass payback of 1,000. If it costs $1,000 per kg to bring shielding materials up from Earth, a mass driver operating from the Moon could reduce that figure to only $1 per kg by the end of construction. When O'Neill figured in all other factors (engineering and logistics), the mass payback cut projected construction costs by 70%.

As the colonists built more colonies, they could try different architectures and different social experiments. They could unburden Earth of its polluting industries by moving them into the vacuum of space, never to corrupt the home world or the interiors of the colonies.

These concepts were popular with techies and many post-60's hippies alike. Here at last was the cosmic hope of the human race, a new frontier to conquer without any natives to displace. A chance to move polluting and extractive industries off Earth forever. A greener future for everybody.

The dreamers within NASA warmed to this vision, so the Agency commissioned cross-disciplinary summer studies at Stanford University in 1975, and at the nearby Ames Research Center in 1977. There was just the matter of cost: optimistically, a hundred times the NASA budget. This far transcended the resources of the congressional VA-HUD committees. To garner that kind of funding, O'Neill astutely realized that space colonies would have to solve major global problems.

At the time, the United States was in the midst of an energy price crisis, and people were starting to perceive that the world had been in a state of perpetual, low-level warfare for energy resources for decades.

Energy is plentiful in space. The Sun shines 24 hours a day, and it carries over twice as much power per square meter than what reaches us on the ground in the daytime. By the 1970s, satellites had been using solar power for years.

What if giant satellites could capture solar power and deliver it to Earth? Several studies suggested that satellites could efficiently beam power in the form of microwaves down to giant Earth-side antenna farms. Some studies even suggested this could provide electricity at competitive prices, if the cost of launch fell significantly.

Here was the answer! Unlimited power from space, anywhere in the world, made 70% more economical by using lunar materials. Because solar power satellites needed enormous collecting areas, they would require thousands of workers to build. The workers, in turn, would need somewhere to live: space colonies.

So the key to selling space colonies, which is what the space dreamers really wanted, was selling solar power satellites. Thus space solar power became the centerpiece of space colony advocacy for decades to come.

However, in the years since O'Neill's work, space transportation and infrastructure costs have remained nearly constant at values much too high for economical space solar power. Meanwhile, as oil prices and world energy demand continue to rise, ground-based renewable alternatives are becoming cheaper. Amorphous silicon solar cells have dropped by a factor of 34 in price per watt.[49] Poor rural areas everywhere are finding it cheaper to electrify by mounting solar panels to roofs than to pay for long stretches of power lines.[177] Passive solar architecture, solar water heating, energy efficient devices and smaller, better insulated houses can reduce home power demand by factors of two to ten. Wind power is quickly becoming competitive with oil, gas, coal and nuclear power.[116] The energy infrastructure shows signs of becoming more diverse, accessible, cheap and decentralized.[178] It's moving in exactly the opposite direction as space solar power. The economic basis for O'Neill colonies, never strong in the first place, has eroded past any reasonable hope of resuscitation.

Colonies and Choices

In recent years, NASA has re-dedicated itself to the original von Braun plan of establishing bases on the Moon and Mars. Many hope this will lead to permanent colonies on these worlds. Indeed, a number of popular books have romanticized a colonial future beyond Earth. Kim Stanley Robinson's colorful Mars novels richly depict the once-red planet's conquest by multiple Earth cultures. Robert Zubrin has presented a very original and much cheaper alternative to the von Braun plan to visit and colonize Mars.

Whether from the Moon, Mars, the NEOs or O'Neill colonies, space sings out to techno-humanists (the kind who read *Wired, Make* and *Slashdot*) as a sort of libertarian utopia; a place to live free of Earth's regressive and constricting ways. O'Neill and others spoke often about the opportunities for diversity and experimentation that space habitats would provide.

If history is any guide, though, building cosmic utopias on the colonial model would automatically and indefinitely postpone those freedoms. Using O'Neill's optimistic cost estimates, Freeman Dyson has calculated that home prices in a space colony would come to some 1,500 times the annual wage that workers living there could expect to earn on Earth.[163] Even if their work provided extraordinary value through such exotica as solar power, rare metals or space manufacturing, the depth of their debt would greatly exceed all historical precedent. By comparison, Dyson points out that the cost of the Mayflower expedition worked out to a mere 7.5 years of wages per pilgrim: a pittance compared to the cost of a space colony.

How would a space colony's financiers secure their investment? How would they prevent misuse of continent-scale power generators or hypervelocity debris? To regulate and control such intimidating assets, surely they would bind the colonists more tightly to Earth than any soldier to any army. By far the easiest way to keep the colonists loyal would be to make them dependent on Earth for their survival. This would require nothing so artificial as a salt tax, just incomplete design. O'Neill colonies would not by the slightest stretch of imagination be designed for self-sufficiency, nor would their builders have any incentive to allow them to gain that capacity.

Thus O'Neill colonies, by their very nature, would fail to live up to the hopes of over a century of space dreamers. Because of their enormous expense and need to make a profit, the colonies offer less freedom, not more, to those who would live in them. Because of their built-in dependence on Earth, they could not serve as arks for humans or wildlife should anything go wrong on Earth.

Planetary surface colonies face all of these problems and more. On the Moon and Mars, key resources such as water are scarce and concentrated in a few locations, inviting conflicts over their access and use. The other planets are even less accessible and hospitable. O'Neill was right: planets are no place for any grand future in space. Neither are space colonies. What's left?

The solar system is strewn with millions of primitive asteroids and comet cores, many of which have all the elements needed to build complete ecosystems. Over a hundred of these require less propellant to reach than the Moon. Because NEOs are neither rare nor centralized, using them to build space habitats would bypass much of the strife built into the planetary conquest that we are embarked on today. But even with the streamlined approach to design and construction outlined in the next two chapters, building gaiomes would still cost a lot. If we depend on them for profit, the incentives of colonialism will assert themselves and subvert their sustainability.

Whoever brings gaiomes into being must transcend today's globalist empires and their quarterly balance sheets. They must set aside even those fears and longings that launched the fleets of Europe and eventually put astronauts on the Moon. They must be organized, yet in harmony with nature. They must be locally self-sufficient, yet globally informed. They would do well to consider new worlds from new points of view: as art, as children, as new life forms with their own innate integrity and value. World-builders must do many things differently, starting with design.

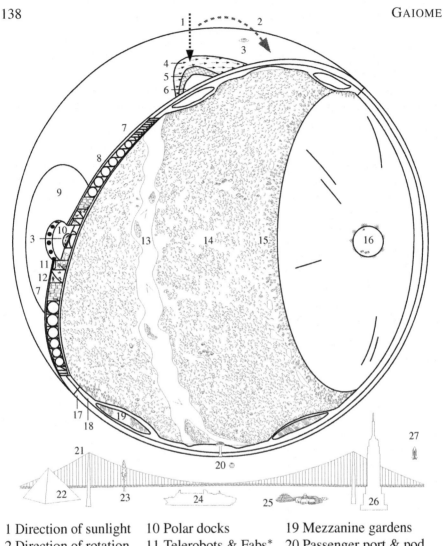

1 Direction of sunlight 10 Polar docks   19 Mezzanine gardens
2 Direction of rotation 11 Telerobots & Fabs*   20 Passenger port & pod
3 Ion Thruster   12 Recreation rings*   21 Golden Gate Bridge
4 Light conduit   13 Equatorial lake   22 Pyramid of Khufu
5 Transceiver mesh   14 Midlands   23 World's tallest tree
6 Structural hull   15 Highlands   24 The *Queen Mary 2*
7 Industrial rings*   16 Sun Image   25 Biosphere 2
8 Propellant storage*   17 Shielding*   26 Empire State building
9 Heat radiator   18 Soil*   27 Space Shuttle

*Thickness and floor height exaggerated 5× for clarity

Figure 5.1: A 2-km spherical gaiome with terrestrial landmarks for scale.

Chapter 5

Design

"Every component of a design should function in many ways. Every essential function should be supported by many components."
—Bill Mollison[151]

You are standing inside a giant sphere filled with sunlight and greenery. A narrow lake shimmers nearby. Through binoculars, your eyes follow it outward and upward until, directly overhead, it squiggles among trees that seem to dangle upside-down. There is no sky. Instead, up there, barely discernible with distance, children are playing with a ball. They kick it toward you, their faces briefly visible, but it quickly loops back to the ground at their feet. Now all you can see is the tops of their heads. The curved floor feels as solid as Earth itself. Yet ten meters below your feet, beneath layers of living soil and luminous hull, it abruptly gives way to the whirling void...

At once familiar and profoundly strange, a gaiome, by definition, would use natural ecology to sustain itself—and you. Like any design, though, it must strike a compromise with the constraints of location and available materials. Thus its verdant landscapes would inevitably wrap themselves around vistas and sensations that simply do not exist on Earth.

As a world in its own right, a gaiome must provide gravity, air, light, water, power, climate control and radiation shielding. To do these tasks efficiently, its design applies the permaculture heuristics we encountered in Chapter 3. These allow it to rely far more on sunlight, geometry and living ecosystems than on heavy machinery to maintain a safe internal climate. Of course the gaiome would still need sophisticated technology to move, communicate, repair and perhaps reproduce. But because we are building for more than human needs, the interiors (Figure 5.1) would look and work much more like forests than like the International Space Station.

This chapter does not so much present a design as provide a guide to designing gaiomes. Just to show what's possible, I give specific numbers for a wide range of shapes, sizes and systems. But the main lessons here are how to consider the entire ecology, how to make each component robust in its own right, and how these systems support each other to make a living world.

Gravity and Architecture

As a hollow object thousands of times smaller than Earth, a gaiome would have no discernible natural gravity. Bones would waste away at an alarmingly rapid rate in this environment, as decades of space flight have verified. To prevent this and other crippling effects, a gaiome would simulate Earth's mighty pull by spinning. The centripetal acceleration so generated would press soils, structures and residents against the inner hull like passengers on a whirling amusement park ride. Would the *pseudogravity* so generated feel natural? Not entirely. When the United States Air Force studied pilots in centrifuges, it found that some began to experience vertigo when spun faster than two revolutions per minute.[179] No one could adapt to six revolutions per minute. On the other hand, no one had trouble at one revolution per minute, so we'll adopt 60 seconds as the minimum acceptable rotation period for a gaiome. The amount of pseudogravity depends on both rotation period and diameter; to provide Earth-like conditions along its equator, a gaiome rotating once a minute must measure at least 1,800 meters across.[180]

You may live *on* Earth, but you would live *in* a gaiome. From the inner surface of the simplest shape—a spinning sphere—"up" would point inward, toward the axis of rotation and "down" would point through the floor, outward into space. If you walked North or South away from the equator toward one of the poles, you would find yourself walking up a hill of increasing slope. Terrace walls that keep dwellings and other structures level would appear and start to get taller. Eventually, at a point about half-way to the pole, you would end up climbing upward more than forward. You would also weigh half as much as you do on Earth, and be able to jump twice as high. Beyond this point, gardens and habitations would yield to low-gravity facilities, air condensers and massive light fixtures.

On Earth, there are three principal directions relative to gravity: Up, Down and Sideways. These correspond to the architectural elements of Roof, Floor and Walls. Sideways motion on Earth is very similar in all directions, so in western architecture, interior walls seldom distinguish between North, South, East or West. That's why, for example, it's easy to get turned around in a large shopping mall.

Architect Theodore Hall has noted that pseudogravity would add two new principal directions to a space habitat.[179] Walking East (in the direction of a gaiome's spin) would amplify the feeling of pseudogravity; walking West would weaken it. Walking North or South would have no discernible effect. It would be very easy to tell directions in a smaller gaiome. Even standing still, you could find directions by simply dropping a small object. Because it would start out closer to the center of the gaiome, its speed would be slightly less than the floor. Therefore it would land slightly short, or West, of the point directly below where you let it go—as if some mysterious force were deflecting it.

Physicists call this fictitious force the coriolis effect. It affects any motion seen from a spinning body. On the slow-spinning Earth, only rockets and long-range artillery need to take it into account. In a gaiome, with its faster rotation, coriolis would be part of daily life. Everything that flies or falls would veer to a greater or lesser degree depending on its speed and direction of motion. Children playing with a ball would quickly discover that it goes farther and higher to the West and that when they send it straight up (toward the center of the gaiome), it will veer to the East.

Orientation and Stability

The gaiome itself would spin along an axis that it would cling to like a gyroscope resisting efforts to knock it over. If we tried to point its North pole toward the Sun, it would stay pointed in that direction, but its orbit would soon bring it around "beside" the Sun. Left alone, a spinning gaiome could only point sunward once a year. We could force it to track the Sun by applying a steady force to it throughout the year. But it is simpler just to let the gaiome spin about an axis parallel to the axis of its orbit about the Sun. In this orientation, the gaiome would appear to "roll" through its orbit (Figure 5.2). Looking

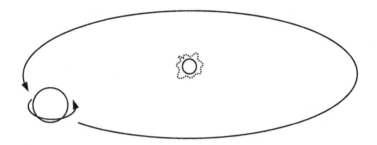

Figure 5.2: Gaiome axes of rotation and orbit.

through an observation window, the Sun would rise and set in exactly the same way every minute or so throughout the year. Of course, the gaiome then would need some way to redirect the light into a steadier, less dizzying path, and I will discuss several methods to do this starting on page 151.

O'Neill's namesake colonies were shaped like cylinders that would spin like drill bits along their long axes. This design came about because as the radius of curvature of a spinning volume increases, the thickness of the hull needed to contain the air pressure (and to a lesser extent, its own weight) rises sharply. Beyond a diameter of 20 kilometers, a colony would need walls several meters thick—an economically impractical amount of steel.[181] The only way left to add area and population to a colony design, then, was to build long and narrow. The spin axis would point toward the Sun and three long mirrors would bring the light into the interior through enormous windows (Figure 5.3).

There are many problems with a tubular design. First, because of its spin, it can only point toward the Sun briefly once a year at most, making sunlight hard to come by. Worse, as sophomore physics students soon learn, long-axis spinners are marginally stable. Perfectly rigid objects set spinning along their long axes should remain in that orientation forever; however real objects such as space colonies are anything but perfectly rigid. Air, water and even people within them would slosh and the mirrors would flex. These motions would gradually shift rotational energy onto the other axes, especially the shortest one, which is dynamically stable. This is the axis that real objects in space eventually find themselves rotating about. America's first satellite, Explorer 1, proved this principle. It arrived on-orbit spinning about its longest axis (the configuration shown in Figure 5.4), but quickly tumbled out of control, ending

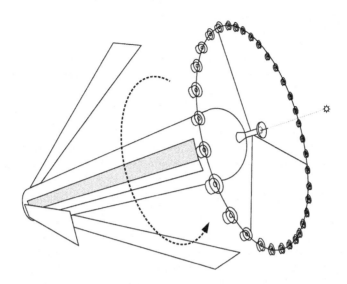

Figure 5.3: O'Neill's Island Two space colony design. Three enormous mirrors reflect sunlight through large windows onto land areas (in gray) inside the cylinder. External agriculture modules appear in the ring on the right.

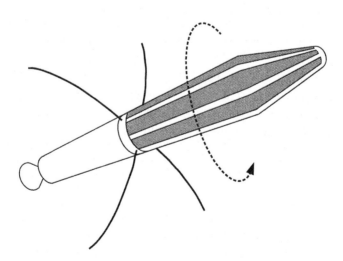

Figure 5.4: Explorer 1.

its first orbit in a flat spin like a juggling club.

To avoid the same fate and keep its mirrors pointed toward the Sun, an O'Neill colony would need to apply corrective forces. This it would do by counter-spinning against another object, typically an identical colony. O'Neill colonies thus would be linked in pairs by massive chain assemblies. Together the twin cylinders would point toward the Sun and spin in opposite directions. But should their enormous mechanical linkages ever fail, both colonies would begin to wobble, eventually tumbling end-over-end, flinging people and objects hundreds of stories along their interiors and perhaps even crashing into each other.

There are other problems with this design. The tips of the mirrors would feel at least 2-4 times as much g-forces as the inhabitants, placing additional strain on the structure. The mirrors could be made thin, but this would reduce their free span: they would tend to buckle under their own weight without lots of additional cabling to hold them in place. More cables, in turn, would require additional capstans to be mounted directly on the windows. All this enormously complicates the problem of shuttering the mirrors for artificial "night."

By contrast, early gaiomes would avoid the risk of uncontrolled tumbling by spinning about their shortest axes. A variety of shapes could accomplish this, the simplest being short cylinders, toroids, dumbbells or (my favorite), slightly flattened (oblate) spheres (Figure 5.5). Any of these shapes is dynamically stable. Should something disturb its rotation and make it wobble, the sloshing air and water in the interior would quickly dissipate the extra motion, returning it to its designed rotation without any extra forces or machinery. However, all of these shapes use a lot more materials per inhabitant than O'Neill's long cylinders. The extra mass would add substantially to the construction costs per unit area. But the added stability, atmosphere, soils and radiation shielding that come with it would also provide a much-needed margin of safety.

Size Constraints

A gaiome's size is constrained on the lower end by human adaptation, as we've seen, and on the upper end by the strength of its construction materials. As was the case for O'Neill colonies, the bigger the gaiome, the greater the strain on its hull due to the pressure and accumulated weight of air, and (to a lesser extent) soil, inhabitants and internal buildings. A two-kilometer spherical gaiome

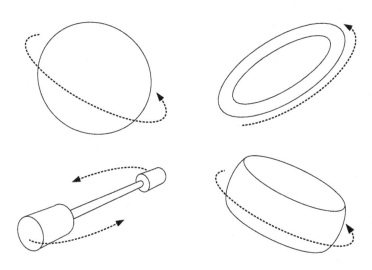

Figure 5.5: Some possible gaiome shapes. Clockwise from top left: oblate spheroid, toroid, short cylinder and dumbbell.

would require a steel hull 54 centimeters thick to hold together; a 20-kilometer gaiome hull would require 5.4 meters of steel.[182]

O'Neill and other designers tried to reduce the mass of the hull (and hence manufacturing costs) by reducing the air pressure. To maintain a breathable amount of oxygen in their habitat's thin air, they chose an oxygen-nitrogen ratio of about 1/1 rather than the 1/4 found on Earth.[181] But that decision only considered animal respiration. The soil microbes that fix nitrogen for plants would be out of luck, making gardening much harder to sustain. A gaiome, on the other hand, would just use a standard Earth atmosphere at full sea-level pressure and rely on stronger construction. To save hull weight, it would use modern composite materials such as woven glass bound in an epoxy or polyester resin matrix (or perhaps simply heated and fused). Structural glass is eight times stronger in tension than steel, and one-third its density, so the hull would have 24 times less mass. Supply matters, too: comets and C-type asteroids may be better-provisioned for glass than for high-tensile steel.

The gaiome's air pressure would decrease with altitude, just as it does on Earth. The air pressure in the center of a 20 kilometer sphere would be half as much as along its equator—too thin to breathe for prolonged periods of time.[183] Spherical gaiomes should not exceed this diameter. However, toroids could be built nearly 300 km in diameter (with a tube diameter of 20 km) before their glass skins reach a meter in thickness.[184]

Supporting a Population

How many people could a gaiome hold? That would depend both on its size (two kilometers, minimum) and how much surface area each person needs—a central question raised in NASA's space habitat studies.[181] After much discussion, the study participants allocated a total of 157 square meters per inhabitant: 49 for housing, 12 for transportation, 44 for intensive plant agriculture, 12 for food processing and drying, and the remaining 40 for all other uses.

While 157 square meters is a pretty small personal "footprint," the studies compared it favorably to the figure of 98 that holds for New York City, or 46 for pastoral Vence, France. This comparison is incomplete, though, because cities and even villages generally depend on distant supplies of air, water, food, energy and construction materials. A city's ecological footprint generally extends far beyond its legal borders. In 1963, when systems ecologist Howard Odum calculated a person's minimum footprint in a permanent space station, he got a figure of one hectare, or 10,000 square meters.[43] At that time, Earth itself had about 4 hectares of dry land per person. Because of population growth, it now has less than two.

Why the factor of 64 discrepancy between NASA and Odum? The NASA study assumed that a space colony would have total control over its agriculture, maximizing sunlight, water, nutrients and the number of growing seasons per year while eliminating pests completely. Indeed, when NASA and the Soviet space programs studied plant growth in closed chambers that met these conditions (pages 90 and 92), they found that they could easily replenish air and water with 20 to 35 square meters of intensive agricultural area per person. Food was more challenging. The Soviets estimated that they could close their food cycle at 56 square meters.[122] But the privately funded Biosphere 2 experiment (page 93), which depended on sunlight and had a third less of it than expected, barely fed everybody with 250 square meters of crop area per person.[133]

Odum's point was that natural ecology vastly surpasses any human life-support technology. Indeed, NASA's "closed" growth chambers were high-maintenance systems with energy and resource footprints extending far beyond their walls. For example, the electrical power required for artificial lighting, plumbing and (especially) temperature control in NASA's test chamber at the Kennedy Space Center was in the 300 kilowatt range.[128] This illuminated 40 square meters of plants. A solar panel big enough to power the test chamber would have to measure over 3,200 square meters. So the chamber's actual footprint was 80 times as large as the growing area it supported. By contrast, Earth life has evolved under intense pressure toward miniaturization, maximum power output, self-regulation, self-repair and resource recovery. In a very literal sense, every cell is a biological experiment, and Earth's research "budget," measured in sheer weight of living beings (biomass), outweighs the human species even today (by a factor of 100,000, according to Criswell).[54] Humanity may have the advantages of intelligence and the abstraction and accumulation of knowledge, but Earth has had nearly a million times longer than written history to optimize its bio-machinery, and the stakes have always been life or death. In Odum's view, it seemed unlikely that a robust life-support system could be made anywhere near as efficient as a natural ecology.

Odum's space ecology would be built up like an aquarium or terrarium, with species added steadily until they achieved balance with each other and their environment. This ecology would include humans from the beginning, bailing them out as needed with infusions of air and food until the emergent system stabilized.[43] At this point, it would support humans without agriculture. In such an Eden-like environment, we could gather and hunt like our ancestors in central Africa did through much of prehistory. But we would need at least 64 times as much land per person as an O'Neill colony.

Food Security

Supporting a dense population is hard to do with Odum's numbers. However, some of Odum's assumptions may not prevail in a gaiome. By design, it would not experience Earth's harsh weather, seasonal extremes, or regional shortages of water and nutrients. Absent these factors, it could enjoy up to a tenfold increase in bioproductivity, thereby supporting a denser population than Odum's model would indicate.[39]

Odum also assumed that meat would provide humans with the bulk of their dietary protein. This places very heavy energy demands on the ecosystem.

For example, the same growing area required to produce a tonne of protein from beef can produce 5 tonnes of protein from grains or 26 tonnes from leafy greens. [185] If even a fifth of your dietary protein comes from animals, it takes two to five times as much land area to feed you as it does a vegan.

The word vegan has been used for over a century to describe people who use no animal products whatsoever—no meat, dairy, fish, leather or even honey. Not long ago, the vegan diet may have seemed extremely exotic to many readers. But in recent years it has hit the mainstream in many industrialized nations, finally shedding its reputation as a "deprivation" diet. Most four-star restaurants now feature vegan dishes, and Amazon.com lists over 250 vegan cookbooks. The American Dietetic association has endorsed vegan diets as "appropriate for all stages of the life cycle, including during pregnancy, lactation, infancy, childhood, and adolescence." [186] Like all diets, it is possible to plan vegan menus that are deficient in various nutrients. However, a diet rich in vegetables, fruits, legumes, whole grains, nuts, seeds and products fortified with calcium and vitamins B12 and D (such as soy milk) is significantly healthier than the average diet of most industrial countries. It is also probably more diverse, as Nyssa Woods dramatically illustrated with a long list of sample vegan menus in her award-winning student paper on space colony design. [187] For economy's sake if nothing else, early gaiomians would be vegan or nearly so. This alone could increase the limiting population density by anywhere from a factor of two to ten.

As ecosystems gain complexity in a gaiome, however, human inhabitants may end up eating the occasional animal. Many plants or portions thereof fulfill essential roles in an ecology without providing any edible food or building materials for humans. As living beings, though, these plants provide food for *something*. The hundreds of niches they would provide may include a few for grazers and foragers such as deer, chickens, squirrels and catfish. A small gaiome may not provide sufficient range for the predators (such as wolves and large cats) that normally keep these populations in check. Thus humans may take on this job by hunting, herding or fishing.

Even so, animals could not comprise a very large fraction of the diet before competing with more efficient vegetable food sources. In Biosphere 2, the animals were chosen specifically not to require dedicated feed crops, but to consume agricultural wastes. They still required a lot of work, daily attention and eventual butchering and processing. Even so, meat, milk and eggs combined to contribute only 13 grams of protein and 150 calories to each person's diet per day—a very small fraction of their needs. [142]

Odum's proposal also concerned ecosystems at climax, that end state of succession seen, for example, in ancient forests. Mature systems have very high cycling ratios and need essentially no management, but many of their yields can be hard to harvest, slow to regenerate, and low in calories. Thus, the defining strategy of gardening and farming is to keep parts of an ecosystem from climax. By laboriously widening forest or prairie margins, farmers make room for early-succession annuals, many of which are starchy, easy to harvest, and store well. Many gaiomians may choose this path, sowing and harvesting in order to enjoy higher population densities, as half the human race has done for nearly ten thousand years. In doing so, however, they need not follow the example of massive, mechanized monocrop agriculture. Instead, they could set aside diverse, interplanted gardens in Zones 1 through 4 around their dwellings, redirecting flows of light, warmth and water through these areas in the correct sequence to maximize yields. Like traditional forage systems that include hundreds of edible plant varieties, the resilient ecologies of these gardens would resist disease, pestilence, floods and fires.

But gardening can only go so far. As human population starts to press up against Odum's limit (2% of total plant productivity: see page 30), our survival depends on exerting ever greater conscious control over the environment. This leaves us increasingly vulnerable to mistakes ranging from simple negligence to economic famine. Higher population density means less food security in any space habitat—even the one called Earth.

Ecological Footprint: A Working Assumption

Population density was a critical parameter for O'Neill because his colony had to make a profit. The more labor he could cram into his colony, the faster the investment would pay off. Gaiomes, on the other hand, would not be built to return a profit, but to express the exuberance of a regeneratively bounteous culture. As living art, they would not require dense populations.

It is helpful, however, to have a concrete number for discussion. While the optimum value won't be known for a long time, all of the gaiome designs in this book allocate 1,000 square meters per person: 100 square meters for intensive gardening, 700 for forest/wild areas (which provide many edible plants) and 200 for living, work and transportation space. While 10 times denser than

Odum's world, it's 6.4 times roomier and far greener than the NASA designs.

At 1,000 square meters per person, a 2,000×100 meter torus would house 750 people, a two kilometer sphere could hold 9,000 people, and a 20 kilometer sphere could support a population of 900,000. The largest shape considered here, a 300×30 km torus, could hold 32 million people. A gaiome consisting of ten such toroids connected edge-to-edge like rings on a finger could support up to 320 million people, though I'm not at all sure it would be to anyone's benefit to build one this size.

Gaiomes potentially span at least five orders of magnitude in population —a daunting range of possibilities to consider. However, some things don't depend on size. For example, a person's minimum ecological footprint (here assumed to be 1,000 m^2) holds throughout the full range of gaiomes because on average people have similar needs for water, light and food, regardless of population size. So it's possible to design gaiomes in the abstract around the needs of the individual, then scale them to whatever size is practical for construction. Let's consider some of those needs now.

Power and Light

Could gaiomes live and run on solar power as Earth's biosphere does? It should be easy: in space, the Sun shines 24 hours a day and twice as bright as it does below Earth's clouds and atmosphere. But how much power would a gaiome need?

According to the U.S. Department of Energy,[188] the average American consumed about 11 kilowatts of power continuously throughout 2003—about twice as much as in 1961, when Yuri Gagarin became the first human in space. This figure includes all industrial, transportation, residential and commercial uses.

Gaiomes could support a more energy-efficient lifestyle. Their small size would eliminate the need for cars, planes or ships, which account for 28% of U.S. power consumption (though the larger gaiomes may find efficient transport such as light rail convenient). By controlling how much light to reflect, absorb and admit into their interiors, gaiomes could maintain comfortable climates, saving another 14% that would have gone to indoor heating and cooling. In the U.S., two thirds of the power used to produce electricity burns up in the generators or dissipates in long-distance power lines. Such electrical losses add up to over 27% of the total power demand. Commercial and residential lighting accounts for another 6%. Industry and commerce within a gaiome may consume less power than on Earth for several reasons. With mild weather, buildings (if needed for storage, ceremony and privacy) would be lighter and need less maintenance. With fresh produce available year-round, food would need less processing and storage. But offsetting these efficiencies, gaiomes would need some additional machinery to supplement their natural life support systems. So let's assume no change to the industrial and commercial power demand. Each gaiomian would then require 2.9 kilowatts.

Now what about light? On Earth, most of it comes for free from the sky. In a gaiome, *all* energy would come for free from the Sun, but it still must be managed and directed. In residential architecture, illuminance (in lux) describes how much light we need per square meter. Each of a gaiomian's 100 square meters of food garden needs about 40,000 lux of illumination during the day. Similarly, each of the 700 square meters of wild plants (including some edibles) needs 20,000 lux.[189] The 200 square meters of living and work spaces might need anywhere between 50 and 1,600 lux, averaging about 500.[190]

Most light bulbs are labeled with their total output in lumens, which is just lux times the number of square meters to be illuminated. Averaged over a 24-hour day/night cycle, the lux values in a gaiome would be half as much as above, so, multiplied by the total area available, each person would require 9,050,000 lumens of light—about 29 times more than NASA's design.[191]

What would it take to actually supply 9 million lumens of light plus 2.9 kilowatts of electrical power from the Sun? A spherical gaiome with a radius of r would catch every solar photon within a circle of area πr^2, exactly one-quarter of its total surface area. But not every part of the gaiome's interior needs direct light. For every 1,000 square meters of inhabited, forested or agricultural land, a spherical gaiome would have 400 more square meters near its poles that would not need direct illumination (indeed, these areas most likely would house the main light sources).[192] That would give each gaiomian a total of $1,400 \div 4 = 350$ square meters of collecting area. Near Earth, the Sun delivers a continuous 1,380 watts per square meter, so the amount of power available is enormous: 483 kilowatts per gaiomian.

This is more than enough to light the interior with electric bulbs. Industry has long used xenon arc lamps to mimic sunlight; at high power (20 kW), a xenon bulb can provide 150 lumens per watt (versus less than 20 lumens per watt for an incandescent bulb). To provide for every gaiomian at 1 AU from the Sun, only 12.5% of the incoming light needs to be converted to electricity. Even relatively cheap photovoltaic solar cells routinely accomplish this today. Current research is driving solar cells toward efficiencies of up to 50%.[193] Light emitting diodes with efficiencies of 300 lumens per watt may one day replace xenon arc lamps. Together these improvements could extend the range of a gaiome as far as 2.8 AU—well into the asteroid belt—where the Sun appears eight times fainter than it does near Earth.

But this approach smacks of brute force. Bulbs burn out. Solar cells degrade in a few decades. Both would require constant replacement. Wouldn't it just be simpler to let the light in directly, bypassing the inefficiency of electrical conversion?

Sunlight produces 4.6 times more illumination per watt than a xenon lamp. But not all of the sunlight arrives in a useful form. Photosynthesis uses light with wavelengths between 0.4 and 0.7 microns, about the same range as our visual perception. Only about 38% of solar photons fall within this range. Earth's atmosphere screens out all but a beneficial trickle of the remaining, invisible photons. This is good news for Earth life, because ultraviolet photons with wavelengths around 0.28 microns destroy living cells on contact.

The surface of a gaiome must adopt the atmosphere's protective role and reject 62% of the incident 483 kilowatts, leaving 180 kilowatts of high-quality sunlight per person. To provide 9 million lumens, only 7.2% of this would need to reach the interior.

Getting the light to the inside presents a problem. For example, allowing for a 50% loss of light in transmission through glass, it would take 23,000 windows 10 meters in diameter to light the two-kilometer gaiome. So many windows would greatly increase the chance of leaks.

Perhaps it's possible to get light inside without using windows. Recall from page 145 that gaiome hulls could use a tight weave of glass fibers for strength. The resins used to bind these into a composite could be designed to trap some of the visible light. Tapered optical fibers embedded in the weave could then bring light into the interior and deliver it where needed. Because any given fiber on the spinning gaiome would cycle between light and darkness every minute or so, it eventually would have to combine with fibers from other parts of the gaiome to provide a steady light source. No problem: good quality optical fibers have essentially no light loss over a few kilometer run. The only part that's unclear at this point is what fraction of the incident light they can capture. Anything over 7.2% would work at 1 AU from the Sun.

What about electrical power? Perhaps some of the unusable ultraviolet light could be converted into electricity. Ultraviolet photocells already surpass the 2% efficiency needed to supply each gaiomian with 2.9 kilowatts near Earth. If ultraviolet-converting coatings become very efficient, interior lamps could extend a gaiome's range farther from the Sun.

Toroidal gaiomes would have habitable areas ranging from two to three times their solar cross sections, which compares favorably to the ratio of 2.86 in the spherical example above. Toroids would have interiors as bright to slightly brighter than spherical gaiomes built with the same light-gathering materials.

Parceling the Light

Sunlight would fall constantly on a gaiome, but with the exception of some grains such as wheat, the plants and animals within would need a daily cycle of light and darkness. Fruits such as tomatoes prefer intense daylight two to four times brighter than the average values mentioned above, interspersed with cooler, truly dark nights. The gaiome therefore must divide the constant light it gathers and parcel it out in a varying daily cycle.

A brute force, high-maintenance approach might be to store the energy and release it in a programmed way. In a spherical gaiome, one could mount large light fixtures inside near the poles and, with a gigantic clockwork mechanism, aim them toward different areas at different times of the day. But the huge amount of scattered and reflected light would do away with nighttime altogether in such a scheme. Another approach would be to shunt sunlight to intensive agriculture areas in separate mezzanine sections (the four white areas of Figure 5.6) during the main floor's night. Under normal circumstances, air would circulate between the sphere and the mezzanine level, though air-tight doors would provide emergency shelter in case of a hull breach.

Toroidal (doughnut-shaped) designs lend themselves to efficient light management. Suppose the light could be routed efficiently into an optical storage ring mounted on the "ceiling" near the toroid's innermost circumference. Light could then be pulled out of the ring via piezo-mounted mirrors that break the light path, or perhaps using solid-state switches to alter the optical properties of segments of the ring itself. Then any part of the gaiome may withdraw light from the ring, creating for itself a bright patch.

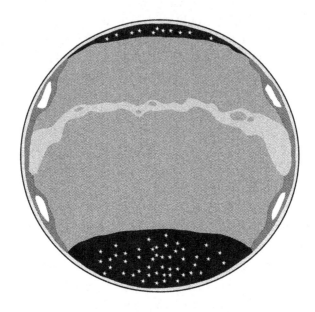

Figure 5.6: Night time in a sphere, with brightly-lit mezzanine levels. The equatorial lake is drawn for reference.

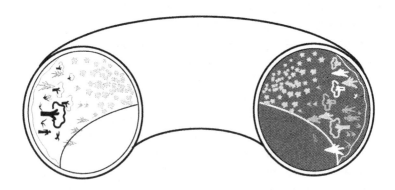

Figure 5.7: Schematic view of day and night in a torus. Not to scale.

A narrow toroid could simulate the Sun's apparent motion through the sky. All it would have to do is rotate its bright patch through a full 360° once per day (Figure 5.7). From any fixed vantage point on the floor of the gaiome, the sky would brighten from the East each morning until the virtual sun rises. After transiting nearly overhead, it would continue Westward, shadows gradually lengthening until lost in twilight.

For truly dark nights, a toroidal gaiome would need large baffles to intercept the reflected light from the day side. Unfortunately, these also would tend to trap flying animals, reduce the lines of sight and increase the sense of confinement for early settlers.

A toroidal gaiome would have extreme time zones. For example, in a gaiome 2 km in diameter, every kilometer would span a four hour time difference. Even in an enormous 40 km gaiome, you could keep up with daylight (or night) at a comfortable walking pace of 5 km per hour.

Of course, all the trouble with baffles and time zones could be avoided by building toroids in pairs attached edge-to-edge like rings on a finger. Then light could be routed first through one torus and then through the other on an alternating cycle.

Spherical designs may lend themselves to a single, gaiome-wide time zone, but lighting them from their poles would not mimic the sun's progress through the sky. Although plants have been grown under stationary light, no one knows how this might affect a complete ecosystem. If the Sun's diurnal motion proves necessary to an ecology, a mechanical, luminous orb could be sent on a daily journey along the central axis of the gaiome. It would simulate the Sun's movement, but this method also just reeks of brute force. Another method might be to use the optical relay system mentioned for toroids to simulate the Sun's path across a tube running from pole to pole. However, it may be sufficient to use this technique to simulate the Sun's motion across the wide polar region, with the opposite pole acting as a backup. In this case, the poles could alternate lighting and cooling tasks seasonally.

Whatever the final geometry, the moving or changeable parts of the lighting system should remain entirely internal, if possible. This would allow the inhabitants to service them directly within the gaiome. For gaiomes near the Sun, this should be easy because there's plenty of light to work with. Gaiomes that go much farther than Mars, however, would need external mirrors to gather sufficient sunlight. These would require advanced technologies to operate and repair reliably in vacuum—technologies that would likely mature gradually as gaiomes are built and moved among the NEOs.

Ecological Design

Thus far we've focused primarily on human needs. What about the plants that support us and the complex ecologies that support them? What are their specific needs and how do we meet them?

Gravity

How much gravity do plants, pollinators and soil microbes need? The answers almost certainly will emerge in due course from experience with simpler systems. If it happens to take 0.7 gees to produce a tomato, no ground-based system can tell you that. The spherical design presented here assumes it takes one gee of pseudogravity to sustain some species, while many can get by with slightly less, and therefore can grow in the highlands. If this assumption does not prove out, gaiomes could not be spherical. However, they could still be made in the shape of a thick, truck tire-shaped cylinder.

Light

Some plants such as wheat thrive in constant light. Others, such as potatoes, do better with day/night cycles. Moreover, the amount of light most conducive to agriculture is uncomfortably bright to most people. Reflected light from food gardens might contribute significantly to the outdoor illumination of other parts of the gaiome, so permanently illuminated areas might be partitioned off from the rest of the gaiome for part of the day.

Electric lighting allows full control of light color and placement. It should be possible to mimic the full range of colors from an indigo dawn to a brilliant yellow noon to golden embers at dusk. Filtered sunlight would offer less color control, but still could vary considerably depending on placement, intensity and weather.

If you look at a clear sky with polarized glasses in the morning or afternoon, you'll notice a dark band half-way across the sky from the Sun (you might have to tilt your head left and right a bit to see it). This is polarization caused by sunlight scattering perpendicular to its direction of travel. Many invertebrates, particularly pollinating bees, see this clearly and use it to find their way beyond about 100 meters from the nest. Fortunately, a spherical gaiome can easily produce polarized light by placing its major outdoor light sources at the poles. As light reflects and scatters from the ground, water features and air, it becomes polarized, albeit to a lesser extent than the skies of Earth. Bees may be able to adapt to this, or it may be necessary to locate bees' nests strategically around the gaiome near the plants they pollinate. If they persist in straying out of range, reflectors, diffusers or even filters can add more polarization to the light as needed, with some loss of efficiency.

Even the relatively short distance between the ground and the light source spells trouble for flying insects, birds and even some bats, who are often led astray by city lights on Earth. In a gaiome, flyers would tend to collide with the artificial sky. However, if the Sun's image could be formed holographically, it would appear to move sideways when approached. Spiraling after a phantom sun, while tiring and stressful, certainly presents a softer selection pressure than sudden impact. A holographic sun would help to maintain populations long enough for the animals to adapt.

Water

Plants transpire water from their leaves into the atmosphere. Spherical gaiomes could reclaim this moisture via condensers mounted to the interior walls near the poles. The natural circulation pattern would move the warm, humid air from the gaiome's equatorial regions through heat exchangers carrying fluid that has cooled in radiators outside the hull. The resultant condensation dripping into the highland reservoirs would require essentially no further treatment. The high location of these reservoirs is important to their dual roles as water towers and radiation shields.

Because the air tends to thin near the axis of rotation, warm, humid air near the equator would tend to rise and cool. This may form a band or tube of clouds along the axis. Under some circumstances, by design or accident, these might even rain. Otherwise, condensation mesh could be suspended at suitable altitudes to provide rain during the local "night."

Soil

There has been much talk of using hydroponics to save weight in space colonies. For simplicity's sake, we'll stick with old-fashioned organic soils, which enlist the aid of thousands of microbial species and insects to keep nutrients in circulation. Thick soils stack many functions to provide a long-term growing medium, a large thermal mass to buffer the hull's hot/cold cycles, resource reclamation services and large reserves of biomass and minerals.

Growing in thick soils may seem worrisome because, as we saw, Biosphere 2's rich soils teemed with microbes that exhaled carbon dioxide faster than the plants could remove it. However, Biosphere 2 was light-starved, so photosynthesis lagged. The gaiome faces the opposite situation. It has 50% more light to work with than Biosphere 2, and can direct it with much greater precision to wherever it is needed. Recall that in more light-rich environments such as the Bios and CELSS experiments, the air and water cycles easily closed well before the food cycle did. By the time the food cycle closes, plant growth could easily end up limited by available carbon dioxide. Under some circumstances, then, gaiomians might welcome the extra respiration that soil microbes provide.

Gaiomes probably won't use the most common terrestrial practices of fertilizing, tilling and spraying pesticides. These tend to scare off earthworms and disrupt the layered ecology of soil microbes, which ultimately lowers cycling ratios and respiration. The resulting nutrient-rich runoff would also require additional machinery and energy to remove and recycle.

So gaiomic agriculture won't resemble Earth farms at all, nor NASA's sterile plastic plant labs. More likely, it will involve something akin to the high-yield garden beds,[194] permaculture and forest gardens[115] that are becoming popular worldwide.

Species Selection

Recall that diversity helps to stabilize Gaian ecosystems. In natural ecosystems, the maximum number of species that an area can support empirically scales with the fourth root of the area involved.[57] A 2 km gaiome with a million times less area than the large continental bioregions could house roughly a 30th as many species, or about 30,000–100,000. Biosphere 2, by comparison, contained 3,800 known species: an appropriate number for its size.

However, if we just throw random species together, we can expect many to die off, leaving behind an ecosystem that may not support us. Recall from page 98 that even with careful planning, Biosphere 2 lost key species, including all of its pollinators.

Clearly, the species we introduce must be compatible with the enclosed environment and each other. As mentioned on page 87, the right mix can increase the cycling ratio of a nutrient by orders of magnitude, greatly reducing the effort required to maintain the system. This will be especially true if the gaiomes include intensive agriculture areas.

Selecting the right admixture and arrangement of tens of thousands of species from a pool of millions may seem a daunting task. It helps enormously that species have already sorted themselves out into associations or *guilds* that work together with the local climates and soils to maintain stable biomes. Gardeners world-wide now use guild lists to build veritable oases in a wide range of environments including deserts and urban areas.[115] Guild lists and the growing expertise of permaculture gardeners will come in handy when planning a gaiome.

As is common in permaculture, a gaiome's first guild members would be selected both to support each other and to provide for human needs. While the guilds become established, they may lack some of the species diversity found in natural ecologies. They also would lack the benefit of mature soils and constant contact with migratory and wind-borne organisms. So to improve the odds of achieving stable ecosystems, we should sow gaiomes with species that show great genetic variation and let their natural variability make the final adjustments to the new soils and climate. This is just how the apple was introduced to North America: grafted varieties died in the new climate and soils, while a few of the many seeds planted had the right combination of traits to survive.[150]

Even with custom climates and fully stabilized guilds, gaiomes would pose an unprecedented challenge: a complete lack of geological processes to recycle certain nutrients such as sulfur and calcium back into the biosphere. In the short run, machinery might take up the slack. In the long run, new varieties or even new species of life would do the job better. These would emerge with time, under selection pressures both incidental and intentional.

O'Neill believed that space colonies could be built to mimic any climate and soil type, opening up the possibility of building orbiting sanctuaries for endangered species. Unfortunately, though, what endangers many species is shrinking habitat, and gaiomes would be small. They also may not be built for decades or even centuries. Time is running out for the California Condor, the Loggerhead Sea Turtle, the Baker's Larkspur and at least 16,000 other cataloged species known to be facing extinction.[195] Guilds that prove stable in a gaiome may never be able to support these animals. If we want to save them, the time and place to do so is right now, here on Earth.

Shielding

Like the skin of most living organisms, a gaiome's hull would have many jobs and many layers. Its outermost layer would hold the pressure of the atmosphere and the weight of interior structures and soils against pseudogravity. It also would conduct power and light into the interior. Its poles, like the poles of any planet, would be cool and help to radiate waste heat away (incidentally providing large cool areas inside to condense water into gravity-fed reservoirs). But its most pressing task would be to provide shielding from the hazards of space.

We live a sheltered life on Earth. Our atmosphere stops most cosmic rays, charged particles from solar flares, ultraviolet photons and meteors, and together with the soil, mitigates the extremes of solar heating from day to night.

By contrast, if you suddenly found yourself without a space suit in orbit, you would pass out from lack of air within 15 seconds—just long enough to get a very severe burn from the Sun's ultraviolet rays. With a temperature-regulated pressure hull around you typical of modern spacecraft, you would be safe from asphyxiation and harmful sunlight, but over time, you would sicken from exposure to solar charged particles and cosmic rays.

The Sun's roiling surface constantly ejects atomic debris such as protons, helium nuclei and the occasional heavy ion. These collectively form the solar wind, which blows outward into the solar system at 400 km/s. Every few years, the Sun's magnetic corona ejects massive amounts of particles into the solar wind in a series of storms that are also commonly accompanied by solar flares. These intensify the wind to radiation levels that would kill an unprotected astronaut in minutes. The particles, with energies of a few MeV (Million electron-Volts) leave ions inside the body's cells. The ions have both short and long-term effects. If there are enough of them (as occurs during a Coronal Mass Ejection), they will break down the machinery of large numbers of cells, causing them to die. While this *prompt* effect is frightening, their *cumulative* effect can be equally devastating in the long run: lower levels of radiation may not threaten a cell's machinery, but they can still cause it to suffer genetic mutations. Over time, these mutations increase the risk of cancer. Fortunately, the outer hull of a gaiome would easily stop solar charged particles.

Cosmic rays are another story. These subatomic charged particles have bounced around in the remnants of supernovae until they reached energies of thousands to (in rare cases) trillions of times more than any particles emitted by the Sun. While there are too few of them in the Solar System to cause any noticeable prompt effects, their cumulative effects can be deadly.

Not only do cosmic rays penetrate much deeper than solar charged particles, they can continue to cause harm after they've been absorbed. When a very high-energy particle collides with an atomic nucleus, the wreckage can include gamma rays and fast-moving nuclear particles. These *secondary* radiation products can do more harm to cells than the original cosmic ray. Choosing the wrong shield thickness can increase the number of secondaries and actually amplify the damage.

Secondaries tend to peak in the first ten to twenty centimeters of most structural materials. The outer hull would absorb most of the primaries, and the lower meter of slag shielding would stop most of the secondaries. The gaiome's interior would layer 5 meters of soil on top of that in the inhabited areas, and a similar thickness of propellants and machinery in the polar regions, reducing the radiation to 0.5% of the background level at Earth's surface.[196]

What about meteoroids? Most of these rocks from space are so small that when they hit Earth's atmosphere at tens of kilometers per second, they harmlessly ablate away and vaporize, leaving brief streaks of light as they do. A few times a year, though, a rock bigger than about one meter will hit the atmosphere and either explode with substantial violence in the air, or make it all the way to the ground.

Meteoroids are uncommon in deep space. Most are so small that a gaiome's hull would stop them with minimal damage. Still, every few years, the gaiome would collide with something large enough to require minor repairs to its hull. However, it would take a rock 100 grams or larger to fully breach the hull and threaten the air supply—an event that might threaten a 2-km gaiome perhaps every thousand years or so.[181] Even then, if the gaiomians patch their hull within a few days, they should lose a negligible fraction of their atmosphere. Gaiome-destroying impacts would happen less than once in a million years.

With vigilant detection and maneuvering, a gaiome should be able to avoid even these serious impacts.

Near Earth, however, the impact hazard poses a much more serious problem. Orbits with altitudes below about 1,000 kilometers encounter enough satellite debris to make living there hazardous: about a 1% chance of destructive impact per year.[197] Granted, Earth's magnetic field shields spacecraft in equatorial LEO orbits from solar charged particles. That will allow gaiomic technology to be tested without lugging up thousands of tons of radiation shielding. But for the growing ranks of prospective space hoteliers, orbital debris is just one of many practical hurdles to overcome before their work can begin to usher in the gaiome era.

Thermal Control

When sunlight reaches the outer skin of a gaiome, a portion of it would reflect away into space. The rest would end up as heat energy inside the gaiome, either after it is converted to electrical power and used, or as it directly lights and warms the inner and outer surfaces.

The gaiome would also lose energy by glowing. Every object in the universe glows, and the color of its glow depends on its temperature. At livable temperatures (around $300\,\text{K}$), the glow peaks in the infrared portion of the spectrum at around ten microns—wavelengths too long for our eyes to see, yet perceptible to our skin as heat. As each point on a gaiome's surface rotates into darkness, it would continue to glow, noticeably warming nearby objects.

One trick to keeping a gaiome at a comfortable temperature is to choose materials with the right combination of reflection, absorption and radiation (glowing) properties. In the living areas, shielding and soils (see below) will provide a thermal mass that keeps the outer skin at a relatively constant temperature throughout its short day/night cycle.

Another trick is to actively circulate cooling fluids through heat exchangers in the interior, or directly around the brightest lights and heat-producing machinery, and then cycle it through thin radiators outside the hull. This is best done near the poles, which are cooler because sunlight arrives at an oblique angle, spreading out over a larger area. There, the radiators would glow and radiate away energy, cooling the fluid before it returns to the gaiome.

The radiators would stiffly resist changes to the interior temperature because cooling is proportional to temperature to the fourth power. For example, increasing the temperature from 300 to 305 Kelvin increases the cooling rate by 6.8%. If the temperature drops to 295 K, the cooling rate falls by 6.5%. This tendency toward a stable temperature could be amplified further by having the rate of fluid exchange depend on the internal temperature. At higher temperatures, cooling fluids might cycle faster or actually change state from liquid to vapor, increasing the radiator's efficiency.

Engineers will recognize in the thermal control system the essential components of an engine: fluids circulating between a heat source and a sink. Turbines placed in this path could compress and heat the fluid to increase its rate of cooling in the radiators, or impede it to provide mechanical power to machinery or electrical generators. Any such machinery, of course, should be made extremely durable, maintainable and quiet.

So far our energy system design has not been very Gaian. Ideally, the gaiome's ecosystems should regulate the efficiency of the radiators depending on how healthy a given temperature is for them. This I will leave as an exercise for the reader!

Facilities and Services

The gaiome's quality of life will depend on its internal facilities and its interfaces with the rest of the universe. Here's a quick survey of its minimum needs.

Plumbing

The first question most school children have for astronauts is, "How do you go to the bathroom in space?" In microgravity, this is a real problem. Solutions have ranged from the now-infamous adult diapers that astronauts have often worn at launch, to the complex system of suction hoses and air jets currently used in the Space Shuttle and International Space Station. But most areas of a gaiome will have gravity, so toilets should work normally.

Except for one small problem: water-flushed toilets make it very difficult and expensive to close the nutrient cycle tightly. Biosphere 2 used a water-based system, which involved collecting all wash water and sewage into a holding tank where bacteria would munch on it awhile, then stirring it and circulating it among fast-growing aquatic plants, which in turn were used for animal feed. At this point the water was purified and the nutrients were concentrated in the animals. By eating the animals, humans could reclaim the nutrients and close the cycle.[133] Similar wetland sewage treatment strategies have been employed in such commercial products as *Living Machines* and *Earthships*.

However, the wetlands approach is a somewhat roundabout and energy-intensive way to proceed, especially if it relies on fish protein to close the resource loops. Is there a more efficient way to proceed?

Perhaps. One thing you'll see at the most progressive eco-lodges is the composting toilet, which in some cases can be a complex device with electric heaters, fans and so forth. However, the simplest case may involve nothing more than a seat mounted sturdily over a large bucket. The procedure is simple: use it just as you would a flush toilet, then scoop shredded dried leaves and rotting sawdust (from a separate container) over the top. This eliminates odors and transmission of pathogens. Then you close the lid. When the bucket gets much more than half-full, it must be emptied onto a compost pile, washed (with the rinse water also going to the compost pile) and replaced. A layer of dried leaves, sawdust, hay or other neutral matter is then thrown over the top of the pile to provide aeration and prevent odors. After a year or two of rotting, the composting process recovers all vital nutrients to produce a soil that can be used directly for food crops.

This may sound horrifically unsanitary, but the thermophilic action of the compost pile eventually destroys all toxins and parasites. Composting was actually invented for sanitation in some parts of India and Asia, where farmers previously used raw sewage as fertilizer.[198] If the chore of taking a full bucket out to the compost heap seems onerous, keep in mind that even flush toilets need frequent and thorough cleaning on about the same schedule.

The reason we don't see many "humanure" compost heaps in modern society is that flooding of incompletely processed piles can lead to serious contamination. However, unlike Gaia, whose large reservoirs can sometimes spill huge amounts of wind and water into a region, gaiomes simply would not keep enough water aloft to supply a heavy flood.

Wash water from laundry, dishes and showers is called greywater because it's not nearly as dirty as sewage (sometimes called blackwater). Greywater can be fed directly to compact marshes, where water plants digest its relatively benign residues. The effluent from this process can then be used for irrigation.

Diameter (km)	Period (s)	Speed (km/hr)
2	63	360
20	200	1,100
300	780	4,400

Table 5.1: Rotational period and speed of objects of various diameter.

To avoid downstream pollution, soaps must be chosen with care; fortunately, commercial greywater-friendly soaps have been around for decades on Earth. In a gaiome, of course, they'll be the norm.

Ports

Gaiomes would have few if any airlocks, which tend to leak. You also won't find many space suits, because space walks require astronauts to pre-breathe oxygen for hours beforehand. This makes it hard to put a worker outside rapidly in an emergency. Instead, Gaiomes would rely on remote-controlled robots to conduct inspections and repairs outside the hull.

Nevertheless, gaiomes would have docking ports for travel pods of various sizes. Unlike air locks, which must pump air into and out of a large volume, ports are basically just air-tight double doors. Pods that dock along the equator would need to match the gaiome's rotational speed (Table 5.1).[199] Once attached, they would dangle at full pseudogravity, so they would require sturdy docking adapters that can bear their weight. They also must time their arrival so precisely that they would need automated guidance systems. Should an equatorial docking maneuver fail, the pod should have adequate propellant to return and dock at the poles.

"Nations" of gaiomes orbiting the Sun together might plan for pod commerce with trip times lasting from a few minutes to no more than an hour or so. They could achieve this safely at separations of several dozen times their diameter.

Gaiomes that anticipate many visitors may need many equatorial docking ports, or they may have a section with many ports at the poles. If that section spins with the gaiome, it will need structures to support all spacecraft docked there against the slight pseudogravity. Maneuvering in this area and the equatorial ports would require fully automatic landing systems with only an abort override. Gaiomes could make life simpler for the spacecraft and de-spin the polar docks, but this probably would add more complexity than the smaller gaiomes would want to manage. Larger gaiomes would have longer rotation periods, so they also may not choose to de-spin their poles.

Communications

Some gaiomes may opt for wireless broad-band for internal communications, others may rely on voice, bells or drums. That's a matter of taste.

External communications are another matter. They should include, at minimum, high-gain radio. Traditional dish-shaped antennae would work, but these create maintenance problems because they need to operate outside the gaiome and point independently of its rotation. Fortunately, the gaiome's large surface area lends itself to another approach: the phased array antenna. Thousands of antennae would be knit into its entire surface. By introducing different amounts of lag into the signals that come in or go out, they combine the wave fronts as if from a giant dish. Such arrays are old hat on Earth: for decades, they've been able to receive or transmit signals from or to multiple locations and frequencies at the same time. Phased array antennae on even the smallest gaiomes could outperform Earth's largest radio dishes.

Information Technology

At the very minimum, gaiomes would need computer facilities to calculate orbit trajectories, read out thousands of sensors, and use the data to model their climate and energy use. Advanced machine shops would need significant computer resources, as would a library.

No gaiome would be complete without a very large digital library that includes the technical knowledge, practical experience, and wisdom that humanity has laboriously amassed over the past few thousand years. It should include not only all of the Western and Asian classics in numerous translations with commentaries, but as many other books, films and sound recordings as it's possible to store. Just as gaiomes don't turn away from the rich contributions of biological guilds, they should keep the knowledge accumulated by the whole human race alive and well in their cybernetic hearts. By combining in a solar internet with other gaiomes and Earth, they could keep this knowledge up to date.

Transportation

It is easier to plan for transportation in a gaiome than in the settled areas of Earth because available sunlight sets a clear upper limit to population. Therefore, gaiomes can and probably should build in the main elements of their transportation infrastructure from the beginning.

Smaller gaiomes could get by in most situations with walkways and bicycle paths. At a brisk walk, you could circumnavigate a two–kilometer gaiome along its equator in less than an hour. There would be no room or need for cars, ships, trains, airplanes or the infrastructure that supports them.

Larger gaiomes might use light rail or elevator-like systems built on a separate deck beneath the main level. The transport deck would provide easy access to entryways throughout the gaiome without competing for space with dwellings and greenery.

Inhabitants of a very large, high-tech gaiome might have transit doors in their villages or even houses. Upon approaching a door, residents might state their destination and how much they're carrying or how many people are in their party. The doors, perhaps after a brief delay to shuffle in a larger car, if needed, would open to reveal an elevator-like room, with seats if requested. Then, in minutes, it would conduct them to wherever they want to go, opening up in a building, a friend's house, a forest, a low-gravity gym—wherever they want.

Fabrication, Recycling and Repair

On Earth, fabrication equipment often has nothing to do with resource recovery. In a gaiome, the same machinery might handle both processes. For fabrication, a gaiome might use computer-controlled foundries, chemical plants, machine shops and looms. Facilities with the greatest energy demand or risk of toxic accidents might orbit at some distance from the gaiome, with built-in mirrors heating solar furnaces. But as much as possible industry should remain inside. That way, there's more incentive to design recyclable products and use less toxic processes. As Biosphere 2 found, certain plastics and metals retain significant toxins after fabrication. These measurably leached and outgassed into the environment. Indoor production in a sensor-laden environment discourages unsafe and toxic processes that would produce items with residual toxicity.

Some of the machinery that keeps industrial production clean may also reclaim materials that get stuck in other parts of the gaiome's ecosystem. For example, systems that recover minerals in the industrial lab might also reclaim minerals that become trapped in the water system.

Products should use materials that recycle easily. Gaiomes would have plenty of solar power to melt and re-form glass and metal, but organic materials such as bamboo fabric and hemp seed plastic degrade gracefully, automatically helping to keep nutrients in circulation.

Rather than giant monolithic factories, gaiomes may end up using many smaller general-purpose machines. Such "desktop fabs" would take a computer drawing and fabricate precision-machined parts out of metal, glass, plastic or cellulose. Primitive versions of these fabs already exist and have demonstrated their utility even in remote, rural parts of the world.[155,156] With sufficient investment, even small microchip fabs could be developed.

Shelter and Furnishings

The mild climate of most gaiomes would place minimal demands on interior construction. Roofs and walls would have to fend off the occasional rain or prying eye, but never a major storm or earthquake. Structures could be thin, whimsical and elegant. A family could custom-build its own house in a matter of days or weeks from such renewable materials as bamboo, sedge, cob, straw, chord wood or even stone, if available. However, stone, steel or other heavy structures over a couple stories tall would have to be designed in from the start to avoid placing undue stress on the hull. Structures built of lighter-weight material such as wood also should not be built very large in order to minimize the risk of fire.

Interior furnishings, lighting and appliances could comprise anything from simple hammocks and open windows to modern, custom-manufactured furniture, electric lights and appliances. Electrical power, communications and light could come directly from the hull through conduits that rise above the soil at various sites throughout the gaiome.

In a comfortably bounteous ecology, people may find little need for large collections of clothes, tools and gadgets. Without much demand for storage space, dwellings might be small. The true riches of civilization—science, art, literature and music—would not suffer. These never needed much to blossom: just fertile soils and minds.

Mining Equipment

Gaiomes should either store the mining and processing equipment that built them, or have a way to fabricate the needed machinery from on-hand or recoverable materials. Then they can reproduce, perform large repairs, or rebuild after disasters. The necessary equipment might account for about 0.2% of the total mass of the gaiome (see pages 201–209 for details).

Labs and Clinics

Even early gaiomes will need fairly complete laboratory, medical and dental facilities. Industrialized nations have anywhere from 200 to 800 people per medical professional; to meet this standard, the first full spherical gaiome (2 km, population 9,000) would need several dozen doctors, dentists, nurses and medical technicians—enough for some degree of specialization. Their offices, as well as industrial fabs and work spaces, would easily fit in the concentric rings in the hull near the poles.

Recreation

One of the triumphs of O'Neill's colony designs was their thrilling accommodation of the views and sensations unique to space. Island Two's gigantic windows would provide direct views of the Earth, Moon, stars and planets from any point in the colony. In *Colonies in Space*, T. A. Heppenheimer wrote in glowing detail about low-gravity gymnastics and swimming areas near a habitat's axis of rotation.[200] In a similar vein, gaiomes almost certainly would include recreational areas near the hub or poles, elevating sports, dance and theater fully into the third dimension.

Views are another story. The design presented here relies on non-imaging optics or photovoltaic/lamp systems for illumination. It also uses the polar regions for cooling and condensation. Thus, any windows would need to be built specifically for viewing activities.

I have long had my doubts about O'Neill's glass sky that rotates once or twice a minute. Perhaps one could get used to it. Perhaps it would come in handy as a point of reference in ball sports. But in a gaiome, the real problem with windows is that they would let in a blast of sunlight every minute. This might make for an interesting novelty (such as at the bottom of a deep pool, perhaps offset with prisms to stagger the sun-bursts), but in practice, I doubt that windows will play a central role in their architecture.

Alternatively, the optical system that brings in light can be programmed to re-form and de-spin portions of the view, presenting a realistic real-time image of the Sun during the day, and the surrounding cosmos at night.

Gaiomes will abound in opportunities for new forms of expression. As living art, the challenge will be to connect these new aesthetics re-creatively with our terrestrial heritage.

Propulsion

To maintain both their autonomy and their sense of connection with the wider cosmos, gaiomes need ways to move about the solar system. Rockets use non-renewable propellants, which, as we saw on page 50, can dominate mass budgets even for one-way trips. Gaiomes definitely should have high-thrust rockets for quick emergency maneuvers such as dodging a meteoroid (a few meters per second should suffice), but long-distance movement by chemical rocket would require prohibitively massive stores of propellant.

Fortunately, the field of electric propulsion is booming right now. Ion engines use electric or magnetic fields to accelerate charged gas particles to high speeds—often much faster than the exhaust products of chemical reactions. These devices have been used in spacecraft for decades. While liquid Hydrogen and LOX, the best chemical propellant combination in common use, gives a maximum exhaust speed of 4.5 km/s, ion rockets routinely achieve exhaust speeds ten to twenty times higher than that. Not surprisingly, electric propulsion consumes far less propellant for a given ΔV. For example, to reach one of the more accessible asteroids from LEO requires a ΔV of around 4,000 m/s. An ion engine with a 40 km/s exhaust can reach this with a propellant fraction of only 0.1, while Hydrogen-LOX requires a propellant fraction of 0.6.

You don't see ion engines launching cargo from the surface of Earth into orbit because these engines typically weigh several hundred times more than their thrust. They are strictly for use in the vacuum of space, where they can gradually build up speed by thrusting for hours to months on end.

Figure 5.8: Solar sail maneuvers: tacking toward or away from the Sun by adjusting the angle of a solar sail.

Besides propellant, ion engines also require a steady and strong supply of electrical power. Any gaiome that uses ion thrusters will need large external solar panels. To change its speed by a kilometer per second per year, a 2-km gaiome with an ion engine exhaust speed of 60 km/s and 15% efficient photovoltaics would need a pair of square solar panels 25 km across. These would not have to remain attached to the gaiome, but could fly nearby, beaming their power to the ion thrusters via microwaves at 85% efficiency.[181] The arrays and thrusters each would account for about 0.1% of the gaiome's total mass; the propellants would add another 10%.[201]

Long-term, gaiomes may come to use the Sun for a more direct form of propulsion, one that doesn't require rockets or propellants. While not yet tested as a means of propulsion, *Solar sailing* would harness the gentle force of solar radiation pressure, an effect already validated over three decades of spacecraft navigation. Solar sails would consist of large sheets of thin, mirror-like material. Gradually, solar photons (particles of light) bouncing away from the mirrors would change the spacecraft's momentum. For example, a gaiome could fly a large, kite-like mirror at an angle that reflects sunlight in the direction of its orbit. The resulting loss in momentum would cause it to spiral inward toward the Sun—in effect "tacking" against the flow of sunlight (Figure 5.8, left). Likewise, it could angle the kite to reflect light away from its direction of travel, adding momentum in order to spiral out (Figure 5.8, right).

Unfortunately, kites are unwieldy. To attain a ΔV of one kilometer per second per year at 1 AU from the Sun, a 2-kilometer spherical gaiome would need a kite that if square would be 1,000 km wide.[202] The mirrors themselves can be made out of micron-thick aluminum attached to thin, woven panels of carbon fiber. Panels as light as three grams per square meter have been manufactured on Earth. Such panels are amazingly strong, heat-resistant and hold their shape well, even after being folded or rolled.[203] With rigging, a kite might mass 7 tonnes per square kilometer, or about 7% as much as the gaiome it pulls. That leaves room in the design to add more kites, multiplying the gaiome's speed. But to date, no one has any experience handling such huge, thin structures.

As gaiomes master the art of solar sailing, they may experiment with using them to concentrate sunlight on their hulls when they are far from the Sun. In this way, gaiomes could eventually extend their range to 300 AU, well past the Kuiper Belt.

Sensors

Internal sensors would monitor the health and status of power cells, batteries, lights, air, water, soils and the systems that connect them. To keep things simple, all sensors should communicate their readings automatically to devices authorized to poll them, preferably using whatever connection supplies their power (perhaps even electromagnetic induction). We want to avoid the situation that arose in Biosphere 2, where people had to go around with notebooks and read off sensors or call them in to the lab via radio.

External sensors would detect hazards, prospect for asteroids and comets, or simply explore. The phased array infrastructure already in place for communications makes an excellent radar. Nuclear particle detectors would help to monitor cosmic ray bombardment. Telescopes sensitive to visible light, infrared, ultraviolet and perhaps x-rays can image near and distant objects, and determine their surface compositions. However, telescopes pose special problems because they typically must remain pointed at an object long enough to accumulate a significant signal, just as a camera needs a steady mount for long exposures in the dark. A telescope attached to a gaiome would need a very smooth clock drive to counter-rotate and keep it in place. Gaiomes may indeed use a few of these, but for more exacting work, they'll use untethered telescopes that orbit nearby.

Gaiomes that have experience with space telescopes and solar kites could use them in tandem as a sort of giant pin-hole camera similar to the *New Worlds Imager* recently proposed to NASA.[204] Using this technique, space telescopes moving around the focal plane hundreds of thousands of kilometers away could then survey the solar system in vivid detail or even image extrasolar planets.

Size Again - The Nitrogen Problem

Clearly, gaiomes could span a wide range of sizes, shapes, locations, climates and cultures. However, different designs demand different mixtures of building materials. It would be inefficient and unethical to leave a lot of waste behind after construction, so the availability of different materials among the NEOs strongly constrains what kinds of gaiomes will actually get built.

For example, air makes up only 0.4% of the mass of the lightest gaiome in Table 5.2, but 22% of the heaviest.[205] Air is 3/4 nitrogen, which, as we saw on page 130, may be relatively scarce among the NEOs. To get enough nitrogen for even our smallest toroidal gaiome from a C-type asteroid, we must bake the volatiles out of more material than we need to make the soil, shielding and hull. After construction, we're left with a slag heap 340% as massive as the gaiome itself. Larger gaiomes would fare worse, wasting many times their own mass in slag. It's better to build small, and somewhat more efficient to build toroids than spheres.

Comets seem to have the best composition for building gaiomes (Comet Hally, for example, had enough nitrogen to build 2,000×100-meter toroids with no waste). But relatively few comets approach Earth and visiting them involves formidable ΔVs: several tens of kilometers per second. Thus to keep things simple, the first small, torus-shaped gaiomes should probably use asteroids for their source material.

Spherical or large gaiomes, on the other hand, would require dozens of times less material from a comet than they would from an asteroid. With much less cargo to bring back to Earth orbit for processing, it may be worth spending the extra propellant to source the later-generation gaiomes from comets.

To achieve zero waste, gaiomes could increase the fraction devoted to soils, shielding and stored propellants. Whether this results in a more or less mobile gaiome would depend on the exact ratio of solid and volatile components in the source asteroid or comet.

Recently, Donald Yeomans has suggested that C-type asteroids may have more volatiles than we think.[174] If he's right, then the waste figures given here for comets and asteroids could bracket the range of nitrogen fractions we might find among primitive NEOS. Even the observed variation in abundances among

Gaiome Type:	2,000×100m Torus	2 km Sphere	40×2 km Torus	20 km Sphere
Mass ($\times 10^6$t):	12	120	5,900	17,000
Percent of Gaiome Mass				
Air	0.4	4.2	8.1	22
Soil	53	63	44	44
Shield	46	31	38	22
Hull	0.6	1.8	9.9	13
Percent of Gaiome Mass Wasted by Source*				
C Asteroid	340	3,700	7,200	19,000
Comet		52	190	670

*Based on retrieval of Nitrogen only.

Table 5.2: Example of mass and waste accounting for several gaiomes.

carbonaceous chondrite meteorites could improve the waste figures by 30% or more with careful prospecting. Clearly research, luck and tolerance for waste will play major roles in shaping the final design.

So will aesthetics. Toroids may make more efficient use of available materials than spheres, but spheres would have more expansive interior volumes, giving life inside much more room to live and breathe. It's a good bet that eventually, most gaiomes will be small and spherical.

Closing Thoughts

The thirty years since the last serious work on O'Neill colonies have seen great improvements in materials science, computing, communications, ecology and our knowledge of the space environment. Thus it should come as no surprise that the design briefly sketched here is intrinsically more robust, safe, bright, spacious, dynamically stable and independent than earlier work.

This comes, of course, at the cost of much greater mass. For example, NASA's Stanford Torus fit 10,000 people into a 10 million tonne package.[181] By contrast, the 2-km spherical gaiome would require a whopping 120 million tonnes to house 9,000 people (Table 5.2). This large difference in masses, more than an order of magnitude, would not come from high-tech gizmos, but from placing the most advanced known life support technology—life itself, in bona fide ecologies—at the center of the design.

That said, there's still plenty of room for improvement. The systems that supply air, water and thermal control in my designs are only weakly and intermittently Gaian. Hopefully later work will find ways, through better passive architecture, to strengthen these homeostatic qualities. The designs presented here also cry out for a more organic way to distribute the light. I'm uncomfortable with the clockwork machinery and microprocessors that my current lighting architecture would seem to require.

This chapter draws heavily on my science training, which strives to reduce problems to their component parts. While such reductionism is a powerful tool, it is not sufficient for the tasks of ecology. The design's success or failure ultimately rests not only on how well each component meets a need, but on how many other needs it addresses and whether its interactions with neighboring components are favorable.

For example, we have seen how the thick hull can provide many different services: containment, support, shielding, light, power, radar, communications and thermal control. It also can indirectly influence air and water circulation and may feel the effects of air and soil chemistry. By layering and knitting metal, glass and soil together in various combinations, the hull can efficiently incorporate available materials. Even so, it's still only half-designed. To be truly Gaian, it needs even more useful connections to the living parts inside— connections that only time and experience with replicate experiments and real structures can uncover.

Chapter 6

Construction

"Waste anything but time." —Saying popular among Apollo
project engineers
"Waste equals resource." —Basic tenet of ecology.

The previous chapter showed how it's possible to design a wide range of gaiomes that would not break the more obvious laws of physics and ecology. In that sense, they are feasible. But what would it take to actually build one? In some ways, the first full gaiome could be a lot simpler than Biosphere 2. For example, it probably would not have to support a desert biome, and it would never have to deal with unpredictable global weather patterns such as El Niño. Still, it would be over a thousand times larger than Biosphere 2, and it would be in space.

Sending even raw feedstock up from Earth is out of the question. The smallest gaiome, a torus measuring only $2,000 \times 100$ meters, would still have a mass of 12 million tonnes. Even with shipping charges as low as \$550/kg to LEO, it would cost \$6.7 *trillion* just to loft the raw materials into orbit. Even in the short run, world-builders may find it cheaper to figure out how to mine the NEOs.

There is so much to be learned, from ecology to routine access to space, that it makes no sense to try to build a gaiome straightaway. Instead, taking Biosphere 2's cautionary lessons to heart (page 100), the learning and the building should come in stages.

Who Will Build It?

Before anyone can build a gaiome, rockets must become safe and cheap. The field awaits its own Henry Ford and many are applying for the job. Who will get it? Who will open the universe to the great mass of humanity and all the kindoms of life on Earth?

Not NASA

Gaiomes are meant to be robust, serene, organic and autonomous—none of which is NASA's style. Everything NASA does is planned, seemingly to the last detail. Yet somehow, the Agency's operations require rooms full of experts sweating relentless crises—the same image depicted in Collier's Magazine's 1952 illustration of work aboard a "space wheel."[29]

I saw this crisis culture first-hand in my work with NASA's Mars Polar Lander. I was a mere technician who had arrived late in the project to operate the Planetary Society's Mars Microphone experiment. I did not work for NASA. Neither did the Russians, who graciously let the microphone stow away aboard their LiDAR experiment like a barnacle. The cigar-box sized LiDAR was the first foreign instrument aboard a NASA probe, so we were under intense scrutiny. I was therefore startled three days before landing when mission controllers pulled me aside and taught me to edit the daily script file for all activity aboard the spacecraft. This should never have happened: mission rules normally put three levels of software tools and management review between me and that file.

It's the same throughout NASA: just about everything might as well have a "do not touch or change in any way!" sign on it; meanwhile two or three people at various levels in the organization end up teaching you how to change or work around it so that you—and they—can do their jobs. Perhaps the war-like culture of control and crisis at NASA traces back to the Cold War, or maybe even von Braun's harried tenure as a Nazi rocketeer. Whatever its origin, it doesn't lead even remotely in the direction of autonomy—the primary attribute of a gaiome.

Gaiomes, at least during their development phases, would mix poorly with today's government agendas. As autonomous entities, gaiomes would not provide any obvious scientific or economic benefits to the nations that fund them. Without solving any significant social problems or enriching powerful constituents on the time scale of a term of office, gaiomes wouldn't advance political careers. Even if gaiomes someday bring about giant leaps in the evolution of life (as I will suggest in Chapter 7), their builders must look beyond governments to find a benefactor.

Not Corporations

As purely financial investments, gaiomes might work as real estate ventures. Bond issues might fund them, but developing the needed technology to build them will take a long time. Because the required capital investments outstrip the treasury of many nations, gaiomes probably won't attract much venture capital.

Non-profits are also out of the running. Even The Planetary Society, Earth's largest space-related membership organization, failed multiple times to launch a tiny solar sail. Having worked there, I can attest that while they have tonnes of heart, they will never have anything like the money it takes to build a gaiome. Neither will their main competitors, the Mars Society and the National Space Society. Over the years, multiple non-profit clubs with cosmic agendas have struggled for awhile to get off the ground, then fizzled for lack of money and talent.

The super-rich?

Whoever decides to build a gaiome must have lots of money, vision and drive. Because (as we'll soon see) the time frame is so long, they also must be motivated by something other than profit in fungible currency. Two entities qualify: religious organizations and billionaires looking to build a legacy. Thus far, no wealthy religious organization has announced any plans for space. That leaves the rich dreamers. The list of wealthy spacers probably runs into the hundreds, but to date only a few have made any public commitments to their dream.

First, there are the builders. **Paul G. Allen**, co-founder of Microsoft, Inc. financed the X-prize winning *SpaceShipOne*. **Andrew Beal**, who made his fortune in banking, founded Beal Aerospace, a now-defunct rocket company, in the late 1990s. **Jim Benson**, who sold his computer company in 1997, founded SpaceDev, the diversified space services company that made the hybrid rocket motors for *SpaceShipOne*. Amazon.com's **Jeffrey P. Bezos** founded the secretive Seattle-based rocket company *Blue Origin*, which plans to offer the Shepard ride (space altitudes but not orbit) from a space port now under construction in West Texas. **Robert Bigelow**, founder of Budget Suites of America, has budgeted $500 million to fund Bigelow Aerospace's habitat modules. **Richard Branson**, founder of over 150 companies bearing the *Virgin* brand, recently hired *SpaceShipOne* maker Scaled Composites under the banner *Virgin Galactic* to build a sub-orbital passenger spacecraft. **John Carmack**, co-founder of id software, entered X-prize competition with his rocket company Armadillo Aerospace. **Elon Musk**, founder of Zip2 and PayPal,[97] founded *SpaceX* to build a beverage-can orbital rocket.

Next, there are the space adventure travelers. To date, five people have paid the Russian space program a reported $20 million apiece for training and a round-trip to the International Space Station (ISS). **Dennis Tito** (who flew from April 28 to May 6, 2001) is a former aerospace engineer and founder of Wilshire Associates, an investment management firm. **Mark Shuttleworth** (April 25–May 5, 2002), founded Thawte, an internet security firm later sold to VeriSign. **Gregory Olsen** (October 1–11, 2005), is a private investor and former Chairman of Sensors Unlimited, an optoelectronics company. **Anousheh Ansari** (September 18–29, 2006), co-founder and CEO of Telecom Technologies, was also the donor for whom the X-Prize Cup was re-named in 2004. **Charles Simonyi** (April 9–21, 2007), an ex-Microsoft executive, flew while this book was in its final edits.

More privately-funded trips are planned, though the lineup changes often. Would-be space tourists have even included such pop-culture personalities as supermodel Cindy Crawford and singers Lance Bass and Madonna.

While a few rich spacers may just want to play with food in space, most are quite serious about establishing a lasting human presence beyond Earth. None, however, are fools when it comes to money. It's a good bet they won't get richer in space than they could with terrestrial investments. But whoever makes space travel safe and accessible will not soon be forgotten.

Of course the new space moguls will try to defray costs wherever possible. Possible revenue sources include satellite launch services, space adventure travel, on-orbit services (repairing satellites) and media licensing. One might naïvely think that any and all revenue sources will do. But the lessons of countless technology start-ups suggest otherwise. As we saw on page 64, it's foolhardy to attack the center. Any space startup that wants to claim a chunk of the satellite delivery market is asking for trouble.

Gaiomes and their precursors won't attract governments and existing businesses, so they're relatively safe from entrenched competition. Just as IBM was not structured for the nearly non-existent consumer computer market in 1978, so NASA and its contractors are not structured to sell space real estate in 2007.

Robert Bigelow, whom we met on page 59, clearly understands this opportunity. To build the inflatable (or, as Bigelow puts it, "expandable") Nautilus space habitat modules, his self-titled company is developing an ecosystem of small subcontractors. As one who has made his fortune in general contracting, Bigelow insists on having multiple contractors for every category, even if it means paying several different companies to make the same component.[206] Otherwise, he would likely fall victim to the standard aerospace industry practice of underbidding to get a contract and then, after lock-in as prime contractor, massively overrunning the budget.[207]

Bigelow plans to make Nautilus modules available on a fixed-price basis to passengers and private companies, either for outright purchase or rent (he quotes a rate of $7.9 million per passenger for a one week stay.)[197] In short, he's offering space real estate.

If someone will bite the bullet and perform the thousand test launches it will take to develop a safe passenger spacecraft, Bigelow very well could end up building the first precursors to gaiomes.

Maybe you!

The rich dreamers can buy all the resources they need to build profitable passenger spacecraft. But the know-how required to build a gaiome draws from a wider pool of experience. Not only will a gaiome require rocketeers, roboticists, meteoriticists, biologists and perhaps a dozen other specialties, but its designers and builders must live and breathe sustainability. The first generation will need teachers: people like those in Ladakh, Kerala or the New Guinae Highlands who may have no concept of trash. Only as the specialists learn to close the loops in their own lives will they begin to gain the kind of "feel of real things" it takes to make a world.

This kind of knowledge has nothing to do with learned institutes. It lives and grows through daily habit, family custom, village tradition. The only thing it has ever lacked, from the stand-point of an evolving body of knowledge, is a convenient means of sharing, reviewing and replicating innovations among any and all who are interested in learning. In the past, when long-range communications were expensive and slow, these functions required massive hierarchical organization to accomplish, and many indigenous peoples don't care for this. In the present, though, communication and computation are cheap, fast and ubiquitous. In a few short years, for example, wikipedia.org has become the most comprehensive repository of human knowledge in history, accumulating nearly four times as many articles as the *Encyclopædia Britannica* with a comparable level of accuracy.[208] It succeeded not only by enabling everyone to contribute their own specialized knowledge, but also by enlisting everyone, through clear rules, in the maintenance of the organization. In a sense, it is alive, evolving and self-correcting through evolving policies and often lively discussion pages. And it is open to all: if you can find it on the internet, you are already a member.

Open and *peer-to-peer* are two faces of a vast new social movement. Rather than fence off knowledge as "intellectual property," we now have the option to set it free for all to use and improve. Linux, one of the most competitive operating systems in the world, is open-source, meaning that it ships with its source code so that anyone can modify it. Similarly, music file-sharing caught on like wildfire because it connected fans and artists far more efficiently than the bloated "content" industry and its endless procession of gate-keepers ever did.

Gaiomes could benefit enormously from open, peer-to-peer social structures. With the right rules of participation, a gaiomic design could emerge spontaneously through a confluence of interest. As families, villages and space hoteliers share, build, evaluate and discuss actual closed-loop technologies, the fundamentals of a new field will take shape. Just as there's nothing but time and motivation to stop you from contributing to Wikipedia, so there could be nothing to stop you from joining the process of world-building.

The tools are evolving. I've begun a gaiome wiki at gaiome.org. It's a good fit to the present problem of bringing people together to help each other live sustainably. But eventually, as more of us gain confidence in the solar economy, more sophisticated tools will become available. Who knows what they'll look like? Hopefully, they will be focused, transparent, free and open. Proto-gaiomes comprising ground-based or orbiting biospheres with differing degrees of closure may come together like Linux, like dozens of Burning Man convocations lasting days to years. Like life, balancing art and science, gaiomes will take shape through constellations of humanity and nature that our conquest-driven consumer civilization could never imagine.

While it's much too early to identify the organizational details, the physical problems involved in bringing gaiomes into being tend to favor certain paths. Let's examine a likely one now.

A Gaiomic Space Program

The days of missions, count-downs, CapComs, flight suits and contractors in every state are numbered. Nothing hard-wired into NASA needs to carry over into a gaiomic space program, including the concept of a "program."

NASA calls its space excursions "missions," accurately reflecting their short-lived nature. Unlike missions, gaiomes should last indefinitely and nurture whatever destinies and ecologies people can negotiate with their fellow species. Missions have crew and ground control. Gaiomes would have natives and guests.

In the 1980s and 1990s, NASA spent many billions of dollars on paper studies re-designing the space station to accommodate international partners and political whims. Gaiomes won't face these costs. Their builders can pick (or invent) a standard and stick to it.

While gaiomes surely would benefit from NASA's well-documented experience with the space environment, prospective world-builders should avoid emulating the Agency's operations.

Figure 6.1 provides a roadmap for building gaiomes. Activities and technologies are grouped roughly by time (left-to-right) and distance (bottom-to-top). Locations include Earth, LEO, relatively stable "halo" orbits in the Earth-Moon system (page 201) and voyages to the Near-Earth Objects (NEO, top). The pages that follow will describe each of the activities in detail, but the basic story begins with permaculture maturing on Earth while probes survey the asteroids. Passenger spacecraft give rise to space manufacturing, while biospheres gradually take shape on the ground, then in LEO. Eventually, mining vehicles travel to Earth-approaching asteroids, bringing material back to fabrication facilities (fabs) in halo orbit. The fabs manufacture heavier mining vehicles, principally out of steel, and then build copies of themselves. From the slag and saved volatiles, they also build proto-gaiomes and send them down to LEO. The new, heavier miners get proportionally greater amounts of asteroidal material and the loop repeats, growing exponentially. After four iterations (and possibly some visits to comets not shown in the figure), there's enough material near Earth to start building full-fledged gaiomes. These would be seeded

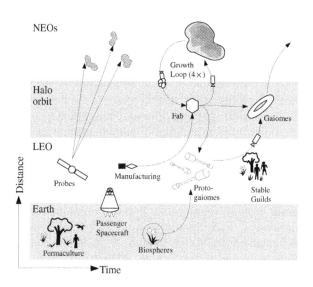

Figure 6.1: Stages of gaiome development at various locations on and beyond Earth.

by guilds that hopefully have had time to stabilize in the proto-gaiomes. Once stable themselves, the gaiomes would visit asteroids, where their built-in fabs could build more gaiomes.

Diagrams similar to this abound in the space community—but with one major difference. In all other development plans, you will find a flow of economic materials or energy back to Earth. Not here. Gaiomes are based on ecological closure, an art best perfected on Earth. Gaiomes can't succeed until Earth's permacultures abound and thrive—at which point, products from space would prove superfluous, if not downright harmful.

Permaculture is the art of living locally and ecologically, an art that took life an enormous amount of time and effort to master on Earth. By re-connecting with natural ecologies, we let go of the task of reinventing the basics of life support. That frees up time and resources to refine our culture, science and ethics. When we build gaiomes, they will be like forest gardens or participatory art—an exuberant expression of our harmonious relationship with nature. Literal vessels of life and dreams, they will exist to flow out into the universe, not to hunt it down and lay it at our feet.

Of course our extractive economy, too, is based on the premise of ever-greater leisure, wealth and art. But its dependence on ever-expanding consumption leads inevitably to the Malthusian agony of resource warfare. Permaculture makes no such demand. It provides generously—so long as the Sun shines and we attend to the circles of life.

Nowhere will we need such attentiveness more than in space. By learning the skills necessary to live permanently on and beyond Earth, we will outgrow the need for imports.

Now to the details.

On the Ground

Before prospective gaiome-builders set up shop, they should take every opportunity to commit to and build a clean, sustainable operation. Ideally, that would mean recycling everything and having no trash whatsoever, using solar and other renewable power sources, and building with non-toxic, renewable materials obtained sustainably (such as FSC-certified wood, plant-based biodegradable plastics and recyclable microchips). Yes, it's expensive. It may seem impossible at first. But building these habits on the ground will cost far less than trying them for the first time in space.

When successful, such an installation would include buildings, soils, water, microclimates, plants, animals, humans and other elements. Together these form a beneficial guild, an entity whose parts complement each other's ecological functions.

The next order of business is building small, sealed terraria that use sterile soils similar to those that might eventually be found among the asteroids and comets. These would help figure out how to get a stable plant biome started with a minimum of chemical or mechanical processing. Success should come reasonably quickly because C-type asteroids and comets are rich in the raw materials of life (see page 129). Unfortunately, pristine carbonaceous meteorites are rare on Earth because they fragment readily during atmospheric entry and decompose rapidly in most environments. But enough is known about their chemistry that it should be possible to construct reasonably good simulants (fake ores) out of terrestrial minerals.

Next would come simple one- or two-person biospheres: basically enclosed guilds. They don't have to be very elaborate, but several of them are necessary to run replicate experiments. Because they would run for prolonged periods of time, their leak rates should be low—10% a year or less. To simulate gaiomes, their interiors need large air volumes relative to their floor space, so they'll have high ceilings. They should supplement windows or opaque walls with sufficient lamps to provide space-like control of light levels.

In this step, clever cost management would really pay off. For example, instead of building proto-gaiomes out of glass, steel and concrete, one could build them out of linked inflatable habitat modules.

Odum has suggested that such closures should start out simple, then add new species steadily. The constant presence of humans would exert a steady pressure on the system to close the air, water and food loops.[43] This does not mean that the same humans must remain cooped up throughout the length of the experiment: crews with similar metabolic demands can and should rotate through on schedules short enough to maintain their health and sanity. Since some or even most succession sequences would probably fail, frequent crew rotation would reduce the stigma of the occasional bail-out.

To help this process, the plumbing should be kept as simple as possible. For example, a humble compost heap might prove more robust than a multistage sewage treatment system. Similarly, suitably gentle soaps would allow wash water to drain directly into garden areas, rather than pass through the wetland filters commonly used to treat greywater.

Where NASA's CELSS experiments used hydroponics, water filters and air conditioning, the closures would use living soils and passive condensation. The experiments should run for as many years as possible, ideally decades. Because of the relatively low cost of these facilities (at least compared to spacecraft), universities and small organizations could operate them. With an open exchange of knowledge among the facility operators, the field would grow rapidly.

In Low-Earth Orbit

During the long interval while permaculture and biospheres mature on the ground, several space technologies should evolve in parallel. The first of these is passenger space travel, which would take shape along the lines I discussed in Chapter 2. Beyond mere access, we'll need several more technologies: automated stationkeeping, docking, telepresence and space manufacturing.

Stationkeeping

Stationkeeping, that is, remaining in a predictable orbit, is essential to long-term safety as the number of satellites and space habitats grows. Staying in a predictable orbit would not be a problem if Earth were a perfect sphere with no atmosphere. In that case, the orbits would be perfect ellipses and the changing position of a satellite could be calculated for all time. But the Earth isn't a perfect sphere: it's somewhat pear-shaped, overlaid with gravitational bumps and wiggles from the continents. Moreover, its outer atmosphere puffs up and shrinks in response to the Sun's rhythms, which vary on time scales lasting days, years and decades. As the outer wisps of atmosphere thicken and thin, the drag on low-orbiting satellites can change by factors of up to a hundred.[209] This combines with the lumpy gravity field of Earth to ensure that orbital paths don't close in perfect ellipses, but become loopy shapes that fill up large volumes of space. There's no fighting some of this drift, so satellite designers work with it to some extent. For example, satellites intended for longer life simply orbit higher, where the atmospheric effects aren't as strong. The pear-shaped gravity perturbation becomes powerful enough at high inclinations that certain orbits actually use it to keep the same angle relative to Earth and the Sun. But even in these situations, there's enough random jitter to require small corrective thrusts to keep the satellites on-station, i.e. in orbits that are easy to calculate. Although these corrections have traditionally been performed remotely by mission control, there's a growing trend to use sensors and software on-board the satellite to make the corrections automatically. A ground observing system may still be needed to verify satellite locations and allocate orbital "slots," but the workload will continue to flow into the satellites themselves.

Docking

Automated docking systems are necessary because (1) space habitats may be built from many expensive modules that need to link up on-orbit, (2) visitors to those habitats would arrive frequently and need to dock reliably and (3) space docking has overwhelmed some of the world's best test pilots on many occasions. Yet the systems involved are simple: compared to aircraft systems that must land in a snowstorm, space maneuvers are easy to simulate accurately. Here again, experience is everything: this technology must mature through thousands of tests to be ready for passenger vehicles.

Telepresence

Space habitats will need external repair and maintenance work from time to time. While donning a space suit may have its romantic appeal, the practicalities are daunting. Space suits take a long time to put on and take off. To preserve the mobility of the arm, leg and finger joints, suits are typically pressurized to a third as much as you would experience inside a spacecraft cabin. Astronauts can't just suit up and get to work at the lower pressure: that would cause nitrogen in their blood to form tiny bubbles, resulting in a malady familiar to SCUBA divers as the bends. To avoid this, space-walkers first must prebreathe oxygen at cabin pressure for several hours to purge the excess nitrogen from their bodies. Beyond the risks they pose to crew, extravehicular activities (astronaut jargon for space walks) also significantly increase the amount of gas that gets vented into space, as well as the risk of spreading debris in orbit.

Telepresence neatly side-steps all of these issues by maintaining a robot outside the spacecraft. Through such devices as video monitors and force-feedback controls, a human operator would control the robot remotely with full sensory feedback. This creates a feeling of being present "inside" the robot that is second only to being there in person. Telerobots significantly speed operations, providing controllers inside the spacecraft or on the ground immediate access to the space environment.

In keeping with the principle of clean operations, robot tools and spacecraft parts should be designed together to ensure that parts and connectors aren't lost during repairs on-orbit.

Telepresence should have a very high priority because it lays the foundation for all large-scale activities in space.

Manufacturing

It is fairly obvious that if the space travel market thrives, demand will emerge for space construction. Passengers will want more comfort and room for their activities aloft. Most passengers would not welcome the sensation of microgravity, which feels exactly like falling, for more than a few minutes. Any longer than that, and even trained fighter pilots begin to experience space sickness, a prolonged, intense vertigo similar to sea sickness. Space sickness is so uncomfortable that many travelers will demand accommodations with pseudogravity. As mentioned earlier, spinning structures can provide it, but must rotate slowly to avoid causing vertigo in their own right. To provide a reasonable amount of pseudogravity, these structures must be much larger than would fit inside the cramped cargo bays of even the world's largest launchers.

Up to a point, of course, structures could be built in several pieces, launched separately and assembled on-orbit. But when a structure requires dozens or even hundreds of launches, design and assembly on-orbit become forbiddingly complex. Then it may be easier just to ship up standard components and assemble them in orbit like the plates of a ship or the boards of a house. This analogy was perfectly clear to von Braun and O'Neill, who imagined space construction projects much like terrestrial shipyards and construction sites: lots of men on-site in protective gear, riveting and welding metal hulls together.

But von Braun and O'Neill missed an important detail: construction sites and shipyards are notoriously messy. On the ground, the debris mostly just sits around as an eyesore. In space, however, a loose bolt always has some relative motion that will take it beyond the construction site. In time, it will end up thousands of kilometers away. Anywhere along its path, it can collide with spacecraft in other orbits at kilometers per second. So to maintain the integrity of the orbital environment, space manufacturing must avoid scattering debris in orbit.

This is hard to do. One could perhaps encapsulate a space construction site inside a gigantic plastic bubble. This would protect the surrounding environment, though the debris it contains would still require careful management during operations, and clean-up when the job was done.

Perhaps construction is the wrong idea. We need fabrication techniques that don't produce debris and that lend themselves to telepresence and automation. Ideally, these methods would take standard ingots of raw materials such as silica and aluminum, and form them without waste into finished hulls, trusses, cables and the like. These are the right technologies for space manufacturing.

In contrast to battleship-era construction, gaiomic manufacturing would lean more toward extrusion, drawing, weaving, knitting, gluing and vapor deposition. These technologies, now fairly mature, allow complex structures to be built directly and seamlessly from simple feedstock. More importantly, they would eliminate such waste products as loose rivets, metal flakes from milling and beads from welding that would permeate a construction site.

Some economy could be realized by perfecting the basic processes in small devices and then using many of them to achieve larger projects. Rather than using a monolithic, custom knitting machine, for example, a project might use ten smaller devices working in tandem. This reaps the benefit not only of lower development, tooling and prototype costs, but also learning curve discounts in manufacturing. True, teaming replicate hardware requires extra logistics, but it's also more flexible and scalable: plant and animal cells use the same principle to achieve a tremendous range of form and function.

The raw materials shipped aloft for manufacturing would come in standard sizes, shapes and compositions. Only the program followed by the machinery would change. Just as a laser printer can print a million different pages using standard toner cartridges and paper, space manufacturing facilities—given time and feedstock—could produce hulls of any desired size and shape.

Space manufacturing, especially with molten glass and metal, would require enormous inputs of power. Fortunately, the Sun supplies 1,380 watts per square meter on-orbit. Where heat is needed, large, thin mirrors could focus sunlight directly with efficiencies well above 50%. Where electricity is needed, photovoltaic panels can provide it with 10-25% efficiency.

Orbiting Proto-gaiomes

With the emergence of large, space-manufactured hulls spun for pseudogravity, the way would at last be clear to develop proto-gaiomes in orbit. The simplest design, resembling a dumbbell, would have two nodes that could run replicate experiments. These would test the effects of low gravity, coriolis and climate on people, plants, soils and animals using guilds from the more promising ground-based closures. Multi-node modules could test numerous proto-gaiomes at once, accelerating the pace of learning.

Because they would orbit below the protective Van Allen Belts, the proto-gaiomes could skimp on radiation shielding, drastically reducing their launch costs. Table 6.1 lists the specifications for a proto-gaiome configured as a space hotel with Mars-like pseudogravity. While cramped compared to the gaiome designs of the previous chapter, it would offer travelers far more room than anything presently in orbit, at less than a tenth the going rate for space accommodations. Much larger versions would be built as asteroidal ores become available (see page 207).

Pseudogravity	0.38 g	Total mass	110 t
Rotation Period	59 s	Development Cost	$10,000 M
Node spacing	330 m	Habitats built	100
Nodes / habitat	2	Unit cost	$380 M
Node Diameter	12 m	Launch cost	$58 M
Decks per Node	2	Crew	2
Deck height	3 m	Visitors	7
Area/person	50 m^2	Room Rate/day	$120,000
Avg. Soil Depth	0.1 m	Break-even	18 mo

Table 6.1: Characteristics of a Low Earth Orbit (LEO) habitat with partial closure, configured as a profitable "hotel."

Near-Earth Objects

Long before manufacturing goes beyond the Van Allen belts, we should take the time to catalog and carefully study the asteroids and comets—the NEOs— that pass near Earth.

When profit-making entities consider new land, they ask what they can extract from it: "What are its economic resources? How can we exploit them?" In this sense, modern business is unique in nature.

When volcanoes make new land on Earth, life courts it gently. Wind and waves bring lichens, fungi, algae, bacteria and seeds to the new shore. Migrating birds rest on the young ash flows. When they take off, they may lighten their load with droppings that bring not only phosphates but sometimes seeds. Eventually, here and there in little pockets, hardy grass and scrub begins to grow. Aided by the rich nutrients found in fresh basalt, clinging stubbornly to whatever moisture they can find, these plants and their successors reach up and lay down to form soils and habitats for microbes, insects and other animals adapted for that locale. Life and land merge in a marriage that strengthens and perpetuates itself not merely through a single lifetime, but across innumerable species and generations.

Asteroids are a different story. Unlike any land on Earth, they do not lie along the way to any other living place we know of. No migratory bird or person will come upon them by chance (except when meteor impacts bring bits of asteroids to us). Asteroids lack gravity, air, liquid water and the protection from cosmic and solar radiation that land life needs to survive.

Perhaps in time we could genetically engineer something like Tsiolkovsky's autotrophs. Living on sunlight, these beings might dwell among the asteroids as naturally as seabirds come to rest on new-formed islands. But as products of our laboratories, the autotrophs would experience a perpetual loneliness practically unknown on Earth. After all, it is unlikely that we could deliberately re-engineer the thousands of species that comprise even the smallest ecosystems that support us, let alone adapt them to open space. Without complimentary species, the autotrophs could survive perhaps through immortality or cannibalism. In either case, though they may trace their lineage back to our DNA, they would not be much like us.

So we're left looking for a middle path. We cannot marry the NEOs like life marries land on Earth, because we don't truly understand life support: the know-how resides not in one species but thousands. Nor should we rape the asteroids and comets for whatever parts of them we can sell. That approach leads straight back to Malthus.

A true middle path would combine the human talent for technology with Gaia's capacity to close the resource loops. We would mine the asteroids, an inherently violent act. As minerals older than Earth vanish in a flash, gaiomes would owe their birth to strife—but not their continued existence. Unlike conquerors, who must grow or die, gaiomes, by design, would live as happily and independently in the solar economy as does their mother world.

Gaiomic surveys of the NEOs would therefore differ from purely economic exploration. Life, not lucre, would be their focus.

Ground-Based Surveys

The first step to mining the NEOs is finding them. Fortunately, surveys already under way have found some 4,115 NEOs (as of this writing), including 57 comets and 783 potentially hazardous asteroids, i.e. objects that could potentially hit the Earth some day. Of the large NEOs, surveys have found an estimated 60-80% of the total population to date and expect to be 90% complete by 2010.[172]

Beyond detection and orbit determination, the next step to learning more about these bodies is detailed observation. Radar measurements taken near closest approach can determine an asteroid's approximate size, shape and rotation period (the length of its "day"). Ground-based telescopes with multiband imagers or spectrometers can determine its surface composition and general type (C, S or M; page 122); however, good spectroscopic measurements exist for only a few dozen NEOs to date.

Taking and analyzing spectra is extremely tedious, but it has already been automated for galaxy work. For example, the Sloan Digital Sky Survey has 5-color images of some 180 million objects, and spectra for 849,920 galaxies, quasars, stars and other objects.[210] Compared to galaxies, NEOs may be harder to find and track, but there are far fewer of them to study. A thorough spectroscopic study of the NEOs is long overdue and would not cost that much.

What makes an NEO a good candidate for building gaiomes? For starters, it should be rich in water and the nutrients needed for life. These are most abundant in primitive objects such as comets and C-type asteroids. Because they are two to three magnitudes darker than similar-sized M- and S-type asteroids, the C-types (and comet cores) will be among the last objects identified in the surveys.

Beyond the demands of life, a suitable NEO should contain the materials needed to build high-tensile strength structures such as the gaiome's hull. Fortunately, a wide range of materials would prove suitable. These include iron, silicon, magnesium and aluminum, which together comprise about 40% of the mass of a C-type asteroid, or 20% of a comet. Finished materials could include steel and other metal alloys, various glasses and glass composites and plastics.

While it would be convenient in the long run to obtain all materials from a single NEO, it is likely that early manufacturing attempts will gather material from different objects as flight opportunities present themselves. Thus it will be important to take spectra of a wide variety of objects including metal and stony asteroids.

It seems reasonable to require that the first NEOs visited not cost much more energy to reach than the surface of the Moon, which from LEO requires a ΔV of no more than 6 km/s. To date, some 535 objects have been found that fit this criterion.[211] They should also be large because, tonne for tonne, smaller objects offer more surface area through which important volatiles such as water, carbon dioxide, methane and ammonia can boil off. Judging from their brightness, 839 of the NEOs found thus far are kilometer-sized or larger.[172]

Combining these criteria (large, primitive, potentially hazardous, and easier to reach than the Moon), we should expect to find around 40 suitable candidates among the NEOs.

Probes

The next step is to send probes out to survey and sample the most reachable and chemically interesting NEOs.

Several spacecraft have pioneered the way to future asteroid probes. In 1997, the NEAR Spacecraft flew by main belt asteroid 243 Mathilde (a C-type) en route to its rendezvous in 1998 and 2000 with 433 Eros (an S-type). In 2001, after orbiting and mapping Eros, NEAR landed on it. While on the surface, NEAR used its x-ray/gamma-ray spectrometer to look for specific chemical elements such as carbon, oxygen, iron, silicon and, indirectly, hydrogen.

The first wave of prospecting probes would probably use something similar to the fleet of identical SIMONE probes proposed for the European Space Agency. At $36 million apiece, these 90 kg (dry mass) "micro-probes" would cost far less than NEAR and similar missions,[212] while still packing essential instruments such as an x-ray spectrometer and infrared imager.

Propulsion

When considering propulsion schemes for the probes, we should balance expedient off-the-shelf solutions such as kerosene-oxygen rockets against the future propulsion needs of the gaiomes themselves.

The SIMONE probes would have ΔV requirements of up to 10 km/s, starting from LEO. Chemical rockets would require propellant mass fractions around 95% to deliver this performance, yet SIMONE would get by with 24%. How? It would use ion engines: electrostatic devices that accelerate ions to much greater speeds than chemical rocket exhaust. With large solar panels supplying the electrical power, the engines would deliver a tiny amount of thrust throughout the mission—much smaller than the vehicle's weight on the ground. While useless for reaching space, ion engines perform well in orbit above the atmosphere. There they can slowly but steadily add speed to the spacecraft without drag losses. The penalty for their high efficiency is time: it takes about two years to apply the entire ΔV of 10 km/s. Despite their scant flight history compared with chemical rockets, ion engines performed well aboard NASA's Deep Space 1 technology demonstrator mission, launched in 1998.

Modern ion engines (including SIMONE's) use xenon for propellant. This noble gas has a high atomic weight and, unlike earlier fuel choices such as mercury and cesium, tends not to corrode the engine's electrode grids. Unfortunately, xenon may be exceedingly rare among the NEOs: gaiomes can't expect to find enough of it during mining and manufacturing to use as fuel.

What about other propulsion technologies? Maybe the probes could use solar kites. Because kites never need refueling, they make the most sense for gaiome propulsion in the long run. A SIMONE probe sans ion engine might carry a 25 kg kite that unfolds to 60 meters on a side. That would give it about the same performance as the ion engine. By increasing the sail to 100 meters on a side, it could double its performance, achieving 10 km/s in a single year.[202]

However, kites don't function well by themselves near Earth. Below altitudes of about 1,000 km, atmospheric drag overpowers the pressure of sunlight. Even above that, solar sails would require significant maneuvering to work their way out of Earth's gravitational field. And they are not without adverse environmental effects: as very large mirrors, they risk blinding instruments aboard other Earth satellites and causing intense flashes on the ground. Therefore, if the probes use kite propulsion, they should also use smaller chemical rockets with 60 kg of propellant to kick out of LEO into an escape orbit. The kites would handle the remaining 2–7 km/s needed to reach the asteroid.

The drawback to kites as a technology path is that they must be enormous to move gaiomes about the solar system: hundreds to thousands of kilometers across. Their sheer size raises serious issues of control and accidental illumination. Even just connecting them to a spinning gaiome while keeping them at a constant angle to the Sun would present a serious challenge. That's a problem for later generations; we should not expect early gaiomes to use kites to travel around the solar system.

Fortunately, a new class of electric propulsion device has started to show promise: pulsed inductive thrusters (PITs). These use magnetic pulses rather than electrostatic wire grids to accelerate ions, so corrosion is negligible. PITs can use just about any available gas for propellant, achieving exhaust speeds between 40 and 80 km/s.[213] Because they offer a development path that even the earliest gaiomes could use, PITs should be thoroughly tested on asteroid probes.

Laboratory PITs can handle about 500 watts per kilogram.[214] Larger, high-power models might eventually handle a kilowatt per kilogram—a figure I use to size the thruster and its power supply from here on.

Sample returns

After dozens of asteroids have been surveyed up close, mining tests can begin on the most promising candidates. Asteroid excavation techniques probably could be tested via remote control, but developing the technologies to convert primitive rock into manufacturing feedstock (metals, volatiles and glasses) will likely involve extensive trial and analysis. A whole succession of processing techniques may have to be tried out before any prove workable. The testing will go faster if it can be done up close, in real time.

Unfortunately, even though they're named for their nearness, most NEOs spend the greater part of their time at distances of one to three AU from Earth. The two-way travel time for radio signals at these distances ranges from 16 to 48 minutes. Even a close approach of, say, 0.1 AU gives a 100 second round-trip talk time—far too long for the immediate tactile and visual feedback required for telepresence.

Thus it makes sense to use the mining tests to obtain substantial asteroid samples and bring them close to Earth. Then different types of processing equipment can be brought up to nearby orbits in quick succession and tried out through telepresence. The samples need to be large enough to work on, make mistakes, and try new approaches. But they also must be small enough to carry back to Earth orbit in a reasonable amount of time.

Tugs

As later probes start to include heavy digging and processing equipment, it may prove wise to separate propulsion systems from surface equipment and cargo. A standardized tug would provide a complete transport solution including propulsion, navigation and a relay link between its cargo and Earth. If it uses PITs or other electrical propulsion, it's power source might also power the mine equipment at the NEO. If it uses solar kites (less likely), these could gather and focus sunlight to provide intense heat where needed for industrial applications.

Table 6.2 shows the performance of a 40-tonne PIT tug with an exhaust speed of 60 km/s. Note the trade-off between cargo mass, total ΔV and trip time. With low-mass cargoes, the tug can make quick, high-ΔV trips to the NEOs, or rapid transfers between LEO and orbits closer to the Moon. Only the inbound trajectories with massive loads of asteroid ores would involve long

	Orbital Transfer		
	LEO-NEO	NEO-L_1	L_1-LEO
ΔV (km/s)	12	5	5
Cargo (t)	50	4,000	110
Propellant (t)	29	350	16
Time (days)	14*	180	8

*plus several weeks coast

Table 6.2: Performance of a 40-tonne tug using Pulsed Induction Thrusters (PITs) and a 40-tonne solar array for various orbital transfers.

trips and large amounts of propellant. Because any volatile will suffice, propellants would come from the NEOs themselves (we'll see how on page 199).

The propellant for the trip back to Earth requires about a tenth of the mass of the cargo. In the case of C-type asteroid ores, that works out to a quarter of their native volatiles, leaving a slag still very rich in organics, volatiles and metals.

Detached Solar Arrays

PITs require large power sources. In the solar economy, this power would come from massive solar-electric arrays. Historically, solar cells have been heavy, typically producing 100 watts per kilogram at efficiencies of 12% or so. That's a little heavy for our purposes. Fortunately, power densities closer to 1,000 watts per kilogram at efficiencies around 30% have already been achieved in the laboratory. They are still very expensive to manufacture. But given that consumer-grade cells have fallen a factor of 34 in price since 1974,[49] it seems likely that high-efficiency cells could be made in bulk at reasonable prices by the time they're needed for tugs.

The tug in Table 6.2 requires 40 megawatts of power. With high-efficiency cells, a 40-tonne array would measure 320 meters on a side. This large, lightweight structure has to remain pointed at the Sun. The tug, which is small and dense, would have an easier time maneuvering with its various cargoes if it can be separated from its ungainly power source. So its array would be detached: using thrusters of its own, it would travel alongside the tug, supplying power with a beam of microwaves. Rectennae (i.e. power-receiving antannae) built into the hulls of tugs and mining equipment would convert this power back into electricity with 85% efficiency.[181] Later generations of tugs, much larger than the first, would have the option to use swarms of solar arrays rather than trying to build ever larger custom power supplies.

Surface Operations

Upon reaching the NEO, a tug would maneuver into a synchronous orbit, allowing it to "hover" over a fixed point on the surface. The several tonnes of mining equipment would find their way to that point (or somewhere nearby) using standard sensors, thrusters and docking software. In the NEO's modest gravity, even an un-powered landing would be possible.

Surface gravity scales directly with density and radius. A kilometer-sized NEO with a density perhaps a quarter of Earth's will have gravity so miniscule that if you stood on it, an object would take minutes to fall from your hand to the ground. With an escape speed slower than a baby's crawl, the slightest bump could accidentally bounce a piece of equipment off into space, never to return. Worse, the asteroids and comets that have been imaged by space probes have irregular shapes, so their surface gravity will vary wildly from place to place. Some areas might have a centripetal acceleration close to the escape speed, making work there essentially the same as working in a weightless environment.

Ground-based survey and space probe data suggests that many asteroids and comets are rubble piles, not solid rocks. So it remains to be seen whether it makes more sense for digging equipment to scrape regolith and rocks from the surface, or simply burrow in and use the cohesion of the surrounding material as a surface to push against.

In either case, the low-gravity environment can be treacherous when working with heavy machinery. A machine with a mass of several tonnes would be light enough on an asteroid to balance on a finger. But doing so would require patience: it would take several minutes to lift it overhead. Of course you could push hard and get it there much faster. But then, if you were out in the open, you might suddenly realize, too late, that it's moving faster than the asteroid's escape speed. If you grabbed onto it at this point, it could easily take you with it on a one-way trip into space.

Even under ground, even if you're very gentle, it's very easy to end up in a situation where you've pulled a piece of equipment for, say, 10 meters,

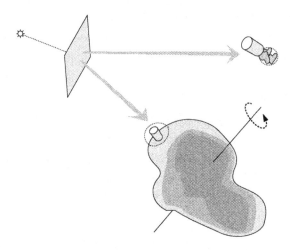

Figure 6.2: The mine site. The tug's detached solar array uses microwave transmitters to beam power to the mining equipment.

but suddenly have only half a meter or so to stop it or change its direction. Now you must apply 20 times as much force, or something—the machine, the surrounding equipment, or even your telerobot—might get damaged.

Small-boat sailors encounter this situation all the time. Boats can coast quite a distance before the water kills their momentum, making them easy to push around, but sometimes hard to stop. Sailors quickly learn to fend off boats in such a way that no part of their bodies gets directly between the boat and whatever they're trying to keep it from hitting. But on or in an asteroid, everything happens in three dimensions, gravity does not help to brace and orient you (or rather, your telerobot), there's no water to slow things down or provide hints of where things are going and it's a long, long, way to the repair shop. So most procedures will require much more caution and formality than we're accustomed to on Earth—especially considering the long signal delays. It will be important to map the surface topography and simulate short sequences of movements prior to uplinking them to the spacecraft, as was done on Mars Pathfinder and the Mars Exploration Rovers.

Because the slow, deliberate excavation requires detailed observation, it will provide excellent data for science and prospecting.

Power for digging and processing will be easy to come by: the solar array, no longer needed for transportation, can refocus its power beams on the mining equipment (Figure 6.2). If it uses a phased array antenna similar to certain radars on Earth, it can split the beam as needed to provide power to multiple pieces of equipment at once.

Obtaining Propellant

As the excavation proceeds, some of the material should be processed to obtain propellants for the return trip. The amount of propellant would depend on the mass of the cargo and the total ΔV required to return it in a timely fashion to Earth orbit, but it will typically amount to 10% of the cargo returned.

For a PIT, nearly any volatile could serve as propellant. A C-type asteroid may contain over 40% volatiles by weight, mostly in hydrated silicates and hydrated salts. These outgas when heated, so our job is to gather some material, provide heat, and trap the gases—mostly water, carbon monoxide and carbon dioxide. A simple bag would suffice to contain these while pumps and cooling equipment liquify and collect each constituent gas. Heat can come from electric heaters or special-purpose solar mirrors. The slag from this process, one-third metallic iron and nickel if baked entirely dry, would be the return cargo.[171] However, we only need to bake out a quarter of the native volatiles for the return trip. The rest we can pull out for various uses in the course of manufacturing in the working orbit near Earth.

Lagrange Points and Halo Orbits

Of course we don't want to bring the asteroid ores all the way down to Earth, use them to build a gaiome, and try to loft it into space: that would be an incredible waste of energy. In practice, we won't even want to bring significant amounts of NEO material as close as LEO. Nudging large rocks Earthward sets a bad precedent. Should a large one go off-course and strike the atmosphere, it might arrive at more than 11 kilometers per second. At that speed, it would most likely break apart and explode.

The explosion might not be small. In 1908, an air burst over Tunguska, Siberia, leveled over 2,000 square kilometers of forest, destroying an estimated 60 million trees. Although there is some debate about the cause, a 10-meter

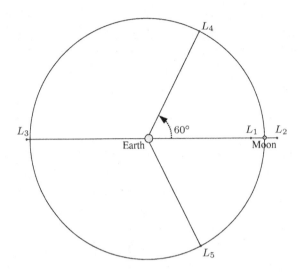

Figure 6.3: The Earth-Moon Lagrange Points.

stony (S-type) meteoroid with a mass of several thousand tonnes could easily have released this amount of energy upon breaking up in the atmosphere.

Any effort to move large asteroid chunks (or gaiomes for that matter) near Earth would therefore be carefully monitored by multiple third parties on Earth. To assuage their qualms, the incoming object should be parked well away from Earth in an orbit that is either very stable or else decays safely away from the Earth. It also should not take too much time or ΔV to reach from LEO or the asteroid.

Although a nearly limitless number of orbits near Earth might fit the bill, two sets in particular stand out: the Earth-Moon and Sun-Earth Lagrange points.

In 1772, Joseph Louis Lagrange discovered that when two bodies orbit each other, there are five points in space where a third object of negligible mass can orbit at a fixed distance from the other two. In the case of the Earth and Moon, the first Lagrange point lies between them where the gravity of the Moon partially counters that of Earth, allowing an object there to have the same orbital period as the Moon. The second Lagrange point lies on the far side of the Moon, where the gravity of Moon and Earth combine to similar effect. The third point lies opposite Earth from the Moon, and the fourth and fifth lead and follow the Moon in its orbit, respectively (Figure 6.3). Lagrange found that when the two primary bodies differ in mass by a factor of more than 25 (as is the case for the Earth and Moon), these last two points are stable. The grass-roots L5 Society that supported O'Neill's work in the 1970s and 1980s was named for the fifth Earth-Moon Lagrange point (abbreviated L_5). The society vowed to disband in a ceremony to be held inside an O'Neill colony at L_5.

Lagrange points abound in the solar system. Trojan asteroids (page 123) cluster about the Sun-Jupiter L_4 and L_5 points. It is possible that Theia (page 18) began life at the Sun-Earth L_4 or L_5 point, either one of which would have been stable until Theia's gravity grew too large. The Sun-Earth L_3 point has been especially popular among pulp fiction authors because it always lay in the unexplored region behind the Sun. Unfortunately, the in-line orbits (L_1, L_2 and L_3) are unstable: any putative anti-Earth at L_3 would soon wander away, never to return. However, it is possible to maintain a "halo" or ring-shaped orbit about any of the first three Lagrange points with relatively little effort (Figure 6.4).

Even the halo orbits are unstable. With the slightest gravitational jostling from neighboring planets, objects that start in halo orbits spiral away from the Sun-Earth (or Earth-Moon) line. After several months, they end up in or-bit around the Sun, Earth or the Moon. Fortunately, within the donut-shaped boundary of these orbits, it takes very little thrust to correct for these drifts.

With planning, the decay of halo orbits can be used to advantage: some trajectories take little to no energy to enter or leave. This property makes halo orbits a likely destination for the tugs and their loads of ore.

Whichever nearby location finally becomes the "working orbit," communi-cations can proceed rapidly with negligible delay. It takes radio signals about two seconds to make the round-trip to the Earth-Moon L_1 point, and about ten seconds to communicate with the Sun-Earth L_1 and L_2 points. Controllers on the ground or in LEO could manipulate telerobots there in near-real time, much as they do today with deep-sea submersibles.

Integrated Processing

Once the NEO ores arrive in a suitable working orbit, processing and refin-ing tests can begin. Space manufacturing (page 188) requires certain feed-stocks: suitable grades of silica, aluminum, iron and other structural materials; volatiles for life support, reagents, plastics and propellants; various minerals

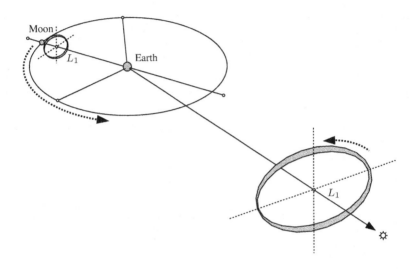

Figure 6.4: Halo orbits about Earth-Moon and Sun-Earth L_1 points.

for shielding and soils. What would it take to make these materials out of raw asteroidal rocks and regolith?

Professor John Lewis has outlined a process whereby successive stages of extraction would leave behind slag that itself could be mined for additional materials. The first probe might extract volatiles and sell them as space propellants. The next might extract the ferrous metals such as iron and nickel and sell them to space manufacturers. Later probes could pull out the platinum-group metals, which could earn handsome profits on the ground. [164,171]

Gaiomes, however, aim to convert NEOs as completely as possible into tiny living worlds, rather than sell off pieces to the highest bidder. But to convert raw ore into a single structure of glass, soil, water, metal and air, all of Lewis's steps and more need to work together in one place.

This sort of integrated processing may fly in the face of the modern tendency toward specialization, but there is precedent. In the 1920s, the Ford Motor Company's River Rouge plant would take in raw materials such as iron and coal ore and work them into finished cars in a matter of hours. [215]

Not only would a single facility fashion a gaiome out of raw ore, but when finished, the gaiome itself should include the equipment it needs to reproduce.

However, unlike a factory, which towers over the cars it builds, the reproductive parts of a gaiome need to be portable: they should account for only a small fraction of its total mass.

This would not be possible if it had to reproduce itself in a day. On Earth, for example, glass foundries typically process less than a tenth of their mass per day.[181] The equipment that smelts, refines, knits and cures the glass hull probably can't beat that figure, especially since it also needs to provide its own heat and electricity from solar mirrors and arrays. The only way to make it portable is to spread the construction out over a much longer period of time—perhaps a year or more.

Let's follow a piece of asteroid ore as it works its way to and through this facility.

Its journey begins with excavation. Terrestrial diggers can be very productive, handling about 500 times their weight per day, but it's anybody's guess how efficient excavation would be in space. Some have guessed productivities as high as 200.[216] I assume a figure of 100. In the low gravity of the NEO, excavation may not require enormous amounts of energy, though there will be plenty available from the now-idle solar arrays that powered the PIT tug.

If the ore is not already regolith, it may be crushed to increase the surface area exposed to chemical processing. Next, solar mirrors would bake it in a sealed chamber or bag, causing it to outgas. Pumps would move, cool and store the volatiles as liquids or compressed gas, as appropriate.

An electric field would sort the dry grains that remain by metal content.[217] Next, carbon monoxide (CO) under just the right amount of heat and pressure would ply the metal-rich grains to form *carbonyls* of various metals.[218]

Carbonyls are moderate-temperature liquids. The carbonyl forms of iron, nickel, cobalt and platinum either decompose to metal + CO, or change from solid to liquid to vapor at very specific pressures between one and 1,000 atmospheres, and temperatures between 30 and 200 °C. This makes it possible to separate metals individually or into various alloys with very little heating.

Metal structures would condense inside pre-formed sectional molds. After pumping in carbonyl vapor of the appropriate alloy, solar concentrators would heat the outside surface, decomposing the carbonyl back to steel and CO. This process requires relatively little energy, but very precise pressure and heat management. Process engineers could take a lesson from terrestrial slip-straw construction and work segment by segment with small molds, using the

Mining or manufacturing activity	Productivity ×equipment mass/day	Power Required	
		Thermal kW/t	Electrical kW/t
Excavation	100		10
Volatile removal	10	140	2
Metal forming	10	14	2
Glass smelting	0.1	1.4	2
Knitting	0.1	0.8	20
Curing	0.1	0.5	2
Soil preparation	100		2

Table 6.3: Equipment productivity (as a multiple of equipment mass) and power requirements (per tonne of material processed) for various space mining and manufacturing activities.

metal formed at each step to complete the seal of the next. In this way, they could build PITs, truss joints, and mining and manufacturing equipment.

The slag that remains would be rich in silica, the main ingredient in glass. Though it's a necessary component of solar panels, optical fibers and composite hull materials, no simple carbonyl process exists to produce glass or fused silica. Instead, chemical baths would leach out the silica, which has a melting temperature of $1,700 - 2,000\,^\circ C$ (additives such as sodium carbonate can bring that down to $1,000\,^\circ C$ or so, but they are too scarce among the asteroids for this purpose). As mentioned before, this step requires heavy handling equipment.

It is much too early to say exactly how big the mining, processing and manufacturing equipment would need to be, nor how much electrical and thermal power it would need. Table 6.3 collects my best guesses, which I use to estimate the total mass of the equipment (see gaiome.com for detailed calculations). Luckily, the most productive equipment would build the largest parts of the 2-km gaiomes. For example, shields and soils account for well over 90% of their total mass, while the glass hulls, which require much more energy and equipment to fabricate, come to just 0.6–1.8% of a gaiome's total mass (Table 5.2, page 174). The facilities a gaiome would need to reproduce itself in a year under these assumptions add up to just 0.2% of its total mass, so these estimates do leave plenty of room for error.

Zero Waste

As the experiment with space manufacturing proceeds, prior success with a zero-waste lifestyle will become more and more important. When every last bit of the asteroidal ore becomes vehicles, habitats and propellants, no debris would remain behind to threaten Earth or space vessels with hypervelocity impacts. This approach would also ensure as complete a marriage as possible between life and the asteroid.

In ordinary engineering, zero waste would be almost unthinkable because the mix of available materials seldom fits the design perfectly. For example, the best solar concentrator mirrors designed to date are made of aluminum and carbon fiber. If these two materials are relatively scarce, achieving zero waste might mean bolstering them with other metals such as magnesium. But since magnesium is less reflective and ductile than aluminum, this would result in heavier mirrors. Similarly, the PITs would have cores made of nearly pure iron rather than stainless steel, which uses relatively scarce chromium. The stainless would be reserved for protective hulls, stress points and some load-bearing structures.

Toxic elements would be removed as early as possible during processing and used strictly outside the living spaces of later habitats and gaiomes. All of the available arsenic, for example, could be bound up with gallium and used in high-efficiency solar arrays, rectennae and transceivers.

The slag left over from industrial fabrication would be rich in silicon, carbon, water and other ingredients found in healthy soil. Modest additional processing would produce a texture and nutrient mix within Earth's wide range of viable primordial soils. Whatever remains would go to shielding.

The extra design work that zero-waste development demands would pay off rapidly due to the mathematical relationship it establishes with the asteroidal materials. The permanent mass payback from the first load of ore could be as much as 50 times the tug's mass. Subsequent trips would multiply this amount.

Growth Loops

The first samples returned to the manufacturing/fabrication facilities (fabs) in working orbit would not have nearly enough mass to build full gaiomes. Thus early efforts would use asteroidal ores to build progressively larger fabs, tugs and mining vehicles. This loop would repeat—a recipe for exponential growth

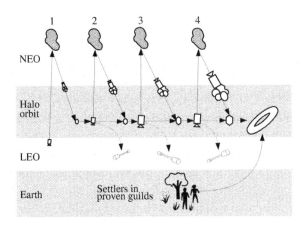

Figure 6.5: Construction timeline.

without significant new shipments from Earth—until both infrastructure and
feedstock grew large enough to build gaiomes (Figure 6.5).

The first tug would carry a small excavator rapidly out to an NEO and
scrape up a load of ore. Then its solar arrays would bake out about a quarter of
the available volatiles (10% of the total ore mass) to use as propellant for the
return trip. Manufacturing facilities waiting in working orbit would process
the cargo, recovering most of the metals and some silica to build larger tugs,
mining equipment and solar arrays. This would use 40% of the ore (Figure 6.6).
Since the tug would return a load fifty times its own mass (including mining
equipment and solar arrays, see Table 6.2, page 196), the second-generation
equipment would have 20 times the mass of the first generation.

Each fleet of miners would probably visit several NEOs as opportunities
arose, rather than trying to accommodate any one object's erratic and infrequent
launch windows. This approach would also broaden our knowledge of NEO
compositions and allow finer adjustment of incoming material compositions to
match manufacturing needs.

After the second fleet departs, the fabs would use the remaining ores to
enlarge themselves and begin building 110-tonne "hotels" of the type described
on page 190. Left-over material, most of it volatiles, would be stored for later
use.

The finished hotels would use the first-generation tugs to spiral safely down

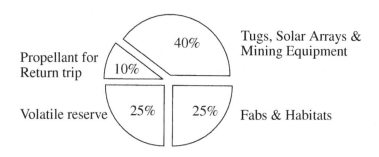

Figure 6.6: Mass breakdown of a C-type asteroid by use during the first four growth loops.

to LEO, where Earth's magnetic field would provide shielding from solar flares. The low orbit would provide safe, convenient access for both tourists and budding ecologies. New species could be introduced rapidly as needed and, should a hotel suffer an accident or a dangerous ecological swing, passengers and crew could bail out and return swiftly to Earth.

By the third loop, operations in working orbit would be large enough to begin building as many as 3 full-gravity 60,000-tonne habitats. A dumbbell-shaped design with two 60 meter diameter nodes 6 decks high could support as many as 34 people with an area of 1,000 m^2 per person. These would house fully closed ecologies, testing the gaiome concept.

But there is a problem: finding enough nitrogen, an essential component of soil and air. Of course it would be nice if at least some of the C-type NEOs turn out to be extinct comets with rich veins of volatiles. However, the only direct evidence to date—meteorite samples such as those discussed on page 130—have perhaps a third as much nitrogen as even the smallest habitats require. This would not pose a problem early on, when each "hotel" needs just 3.8 tonnes of nitrogen. But each of the larger habitats would need 2,200 tonnes, far too much to haul up from Earth.

At this point, it makes sense to begin extracting ammonia and carbon dioxide entirely on-site at the NEO. This process, which mostly involves simple baking under solar concentrators, would provide much-needed nitrogen and carbon to the habitat-building efforts without also junking up the working orbits near Earth with slag.

If these efforts don't turn up enough volatiles, tugs from earlier voyages could make fast trips out to comets or Main Belt asteroids to get more. For example, during the third loop, a second-generation tug could carry some 1,000 tonnes of mining and processing equipment out to a comet. There it would use fairly simple de-volatilization processes to extract 5,700 tonnes of water for propellant and 30,000 tonnes of ammonia (the primary source of nitrogen). Its light loads out and back would allow it to make high-energy trips, achieving an outbound ΔV of 20 km/s in 90 days, and an inbound ΔV of 10 km/s in 550 days. Thus it could reach a large fraction of the comets and C-type asteroids between one and three AU from the Sun, where it would still be close enough to gather enough solar power for engines and operations. However, because the tug would operate too far from Earth for easy telepresence, this approach would seriously challenge the skills built up on the ground in previous voyages.

The fourth loop would bring enough asteroidal and cometary mass to the working orbit to build one or two 2,000×100-meter toroidal gaiomes, each capable of sustaining 750 people. Exponential growth ends when the fabs begin building full gaiomes rather than more mining and manufacturing infrastructure.

If by this time tests with the dumbbell habitats suggest that gaiomes need larger air volumes (e.g. so that birds and insects don't fatally crash into ceilings), a fifth loop would bring enough mass to build two to four 2-km spherical gaiomes, each housing 9,000 people. As mentioned on page 173, spherical gaiomes demand so much more volatiles than small toroids that it may be worth the extra ΔV required to visit comets rather than asteroids in the fifth loop and successive voyages.

Assembling a Gaiome

It would take a year to knit the gaiome's hull, a few weeks to verify its structural integrity, optics and electronics, and a month to spin it up to one full gee of pseudogravity (this would also test its built-in PITs). During spin-up, teleoperated machinery would prep, terrace, grade and chemically balance the soils with nitrogen, carbon, calcium, phosphorus and other nutrients needed for plant growth. Water would condense and flow into an equatorial lake.

With air and bare soil ready, seeds encased in tiny balls of clay and compost from stable guilds would be strewn far and wide. Condensation netting in the high cloud zones would bring rains. The moisture would hydrate and activate the fungi, lichens, soil microbes and insects embedded in the seed balls. Grasses, wildflowers and other pioneer species would then sprout out of them into living soils. These would prep the ground for the next stages of succession, including berry bushes, fruit and nut trees, birds, insects and people. The young gaiome's ecology might take anywhere from years to decades to stabilize, during which time it should stick close to population centers such as Earth and other gaiomes. However, because of their enormous size, the early gaiomes would never drop down to LEO, but would move to high, safe orbits in the Earth-Moon system. Fast tugs could then cycle passengers and habitats between LEO and the gaiomes as needed.

Travel and Reproduction

The whole point of building gaiomes is to establish autonomous living worlds: even the earliest ones should be born mobile and fertile.

The multipurpose mesh knit into a gaiome's hull would serve not only as an antenna for communications and radar, but also as a rectenna to receive microwave power from detached solar arrays hovering nearby. Built-in PITs would use stored volatile propellants to supply it with a ΔV of perhaps 5 km/s for a spiraling journey to an NEO.

After a year or two in transit, a small toroidal gaiome would arrive at the NEO with some 34,000 tonnes of mining and fabrication equipment (about 0.2% of its total mass). This it could use to replenish its propellants in a few short weeks, or to reproduce itself over the course of a year or two.

The industrial capacity of the first gaiomes would match or exceed that of the fabs that built them. By visiting the NEOs directly, gaiomes could operate mines and manufacturing facilities on-site. This would reduce or eliminate the need for long-distance tugs. It would also cut radio lag (which was several seconds between Earth and the halo orbits) down to essentially zero, greatly improving telepresence. Thus, from the second generation forward, most gaiomes would be built far from Earth.

As their ecosystems mature, some daughter gaiomes might remain at the NEO, occasionally taking on settlers and new species as new gaiomes visit. Others would cycle in low-ΔV orbits between Earth's Lagrange points, the NEO, or other destinations. Most, however, would probably cluster together in little groups, forming nations with others that share their language and customs.

The Pull Economy

Establishing the first gaiomes would require a major economic push, the exact size of which is hard to estimate. The uncertainties involved in world-building greatly outnumber those we discussed for tiny passenger spacecraft in Chapter 2. Better access to space surely would make space activities more affordable. So would using indigenous resources, automation and resilient natural life-support systems. But we have a lot to learn in each of these areas, and it's hard to predict how much the lessons will cost. With so many unknowns, gaiomes make no sense whatsoever as a business venture. But as pure art, they could literally take on a life of their own.

Vernadsky wrote of the "pressure of life," that tendency of living organisms to reproduce and push into new areas. "The careful observer can witness this movement of life," he wrote, "and even sense its presence when defending his fields and open spaces against it."[219] But as much as life likes to spread and grow, it also likes company. Pioneer species prepare the way for others to follow. The later-succession species, in turn, often bring new services to the incumbent guilds.

In the vast emptiness of space, this second, associative property of life would exert a tangible pull that would eventually reverse the economics of world building. New gaiomes would find themselves rich in material and energy resources—and short on skills and biodiversity. With plenty of tugs and left-over volatiles to propel them, these tiny worlds would have both motive and means to recruit settlers, offering free passage from LEO outward. Thus their economy would begin to "pull" people and guilds up from Earth.

In the Pull Economy, the best way for starry-eyed youth to find their way into new gaiomes might be to live permaculturally, develop skills such as electronics, forestry, medicine or teleoperations, and save up for a launch to LEO. But what does it mean to live permaculturally? To live up to its billing as permanent culture, the practice must transcend the individual and the household. Gaiome builders would be wise not to recruit individuals or even families. Instead, they should bring in whole villages that have managed to become sustainably self-sufficient in food, water, shelter and clothing.

To avoid a "brain drain" of permacultural know-how from Earth, the space-bound villagers would need a succession plan to teach the new stewards of the land its ways. Rather than simply acquiring it and moving in, buyers would spend a year or two apprenticing with the sellers before launch.

Throughout both push and pull phases, young, autotrophic worlds should take care to germinate fully within cyclic and solar economies rather than trade-based ones. That way, if world-building happens to take longer than expected, it won't especially threaten anyone's life or livelihood. During the push phase, Earth should not import anything from space for profit, but instead donate root stocks, seeds and talents from places that abundantly regenerate them. During the pull phase, gaiomes should never import consumables from Earth for survival, but instead recruit species and skills as needed. Then Vernadsky's pressure would take over within them to fill out their budding ecologies.

Summary

Like the first great cathedrals, the purpose and overall layout of the early gaiomes would take shape long before the technologies necessary to build them have matured. For this reason, learning and accessibility dominate the construction plan. Here are the key steps:

1. Begin with permaculture, the art of living locally and ecologically. Begin on Earth, where the learning is easiest and most needed. Begin now, because the necessary habits and connections take time to establish.

2. Build true passenger spacecraft as soon as possible. This takes space travel out of the realm of conquering elites and makes it a human endeavor.

3. Build many small biospheres, starting with varying degrees of closure on Earth and then in Low-Earth Orbit (LEO). This makes space travel an ecological endeavor.

4. Become skilled at space manufacturing in LEO, using small automation and telepresence systems to work raw feedstock into probes, tugs, habitats, etc. Work clean to preserve the space environment for future generations.

5. Bring asteroid ores from primitive Near-Earth Objects (NEOs) to high, accessible working orbits such as the halo orbits around the Earth-Moon or Earth-Sun L_1 points. Learn how to process this material, then use it to manufacture ever-larger tugs, miners and fabrication facilities (fabs).

6. Manufacture and stabilize gaiomes in working orbits using NEO materials. Building this close to Earth (rather than near the asteroid) allows quicker intervention during the initial assembly of the ecology, and a richer choice of species introductions as it matures.

7. Send stable gaiomes out to the NEOs with mining and processing equipment and populations competent to operate it. Their "pull" economy, with tugs and cycling gaiomes, then opens the solar system to whole villages at the individual cost of a one-way ticket to LEO.

Although I present this plan as a space program, it almost certainly exceeds the abilities of any single entity such as NASA. While government agencies already excel at surveying our neighboring worlds with robot probes, they have proven completely ineffective at building reliable passenger spacecraft. The private sector might succeed if bold investors and entrepreneurs will focus on the passenger market. That could put many more of us into orbit. But private industry's long history of extraction and pollution disqualifies it from building gaiomes. Nor are corporations likely to change: their hierarchical structures and underlying assumptions (such as economies of scale and global branding) don't address permaculture's intensive locality. Thus, before we can begin any significant evolution into space, millions of homes and villages must do their permacultural homework. Only a zero-waste society can build zero-waste governments, companies and space habitats.

In the end, the exact plan and configuration of players won't matter as much as why humans try to live in space in the first place. Design embodies intention. If we intend to escape or conquer, space travel will tax Earth at its time of greatest need and expand our cycles of war and misery. But if it proceeds as part of our efforts to live more harmoniously as ecological, cosmic beings, space travel will open new vistas and challenges for the countless species that comprise our full selves.

Chapter 7

Adaptation

"If you look around at societies to find those that have come into some sort of a peaceful harmonious accommodation with themselves, many of them turn out to be on islands, where the myth of a new frontier vanished long ago."
—Dennis Meadows[3]

In the 1960s, the Russian astronomer Nikolai Kardashev defined cosmic progress in terms of how many worlds, stars and even galaxies a civilization could yoke to the engines of its industry. In his view, civilization would progress through three stages: Type I civilizations would harness the total power output of a planet, Type II, the total power of a star, and Type III an entire galaxy.[220] Kardashev's scheme, basically consumer culture writ large, has long been popular among futurists, exobiologists and science fiction readers.

Yet as we have seen, such dreams of boundless growth and conquest were born in a world long-since transformed by its ecologies into a biological treasure trove. Earth's primary wealth is measured not in precious metals but in chloroplasts, which produce the chemical income (oxygen and ATP) that sustains life. The frontiers we tell each other we've conquered were actually developed and improved over a span of time and effort thousands of times greater than our tenure as a species. Humans never conquered the world. Gaia did, and much of our modern opulence comes from spending down her natural capital.

In space, we find ourselves poorer than we've ever been. To date we have yet to find even a single living cell out there, let alone the vast web of beings that sustain us on Earth. Life will be hard if we try to go it alone. If we try to own the Moon, the Sun or the Galaxy in the same way we have every parcel of Earth, our efforts will forever resemble the disappointments of the Space Shuttle and ISS: expensive, boring and sterile.

Fortunately, as the preceding chapters have shown, we don't have to leave Gaia's love behind. As children of the universe, we can make personal presence our priority in space. As gardeners, we can bring along the seeds of life's regenerative guilds. As architects, we can prepare vessels of gravity, light, soil, water and air to house them. As permaculturists, we can arrange for ever more complex cascades of energy through these living worlds, making them ever richer on a fixed solar income.

Living in space is about world-building, and the world's ecosystems involve cycles of connection, not command. Gaia doesn't own the mineral, liquid and gaseous components of Earth. Her gentle touch co-opts less than a tenth of a percent of the solar power that reaches Earth's atmosphere. Yet that tiny caress has been enough to completely transform the world. Gaia has not enslaved Earth's geology and climate; she has married them. Similarly, to live indefinitely and joyously beyond Earth, Gaia's children must merge with and commit to the lands they find.

If we want a bright future in space, we must abandon Kardashev's scheme of monolithic cosmocracies bent on endless consumption. Instead of asking how large we can grow, let's ask how small we can make our living worlds. In what diversity and multiplicity can they exist? And what would a future teeming with little autonomous worlds mean for humanity, Gaia and life?

This chapter examines this last question on two time scales. First, we'll consider the relatively near future starting a few decades from now and lasting perhaps a page in the Book of Earth. We want to know: Who would inhabit the gaiomes? Why would they go? How might they live? What problems would they face and how would they solve them? Next, we'll turn to the far future, extending out to hundreds of new volumes. Here we'll consider the long-term prospects of gaiomes as life forms in their own right. What might they mean to Gaia and the process of evolution itself?

The Near Future

The first people to say goodbye to blue skies will occupy worlds millions of times smaller and dozens of times less diverse than Earth Gaia. With much smaller reservoirs of air and water, any broken resource flows would become apparent much faster than they do on Earth. For this reason, it would be tempting to conclude that these mini-worlds could serve as analogues to Gaia, alerting us early and often to the consequences of our resource decisions on Earth. But in practice, gaiomes may not share enough of Gaia's mechanics to make such analogies valid. Everything from the path that light takes into a gaiome to the path that heat takes out of it would be different in these inside-out worlds. Without volcanism, plate tectonics, a magnetosphere, an ozone layer and oceanic burial of plankton, the boundaries of a gaiome's basic life systems would be completely different from Earth's. Gaiomic know-how will be an art unto its own, with only incidental parallels to Earth.

Who would live in such a different world? For those who still believe in space colonies and the Final Frontier, the answer seems clear: rugged people wishing to make a better life for themselves, or to escape a bad one. But as we have seen, space is not a porous frontier like the Great Plains of North America. You can't simply walk or canoe across it like Lewis and Clark. Space is far colder than Antarctica, dryer and more desolate than Death Valley. The individual, no matter how brilliant or strong or well-equipped, cannot go beyond the atmosphere and live off the land. Surviving in space requires much more than heroism; more even than human communities at their best and most organized have ever accomplished. It requires thousands of species linked in a Gaian ecology.

The people who come to live in a gaiome had better be very good at nurturing the land wherever they are. They must be able to grow their own food without constantly buying fertilizer, mulch or pesticides. They must be able to house and clothe themselves without importing lumber, cloth or other building materials. They must get water without mining a fossil aquifer. They must never waste, but instead work with the ecology to transform whatever they are finished with to something of equal or greater value. Above all, though, they must make peace with each other and their ecology. Just as permaculture was never just about the individual, so too with gaiomes: they will be settled by communities that, like Dennis Meadows's fabled islanders, "have come to a harmonious accommodation with themselves."

Yet the space community has long been suspicious of societies that harmonize too well. In 1964, Cole and Cox warned that those driven to "conform and adapt" to their circumstances would view space travel as an egoistic abomination.[166] Thirty-six years later, Robert Zubrin echoed this fear, warning that the path that leads to "Unity, Uniformity, Stability" also would lead to stagnation and tyranny.[218]

The more you learn about ecology, the stranger this view seems. Living locally does mean adapting to the immediate environment, but it also means living more autonomously, not less—as we saw with the privatization of Cuba's state farms (page 108). As for adventure, permaculture does not repress it, but merely puts it into the perspective of ecology and culture rather than individual dominance. The first step of permaculture design is adventure itself: you walk the land, noting its boundaries and the flows across it. The Earthrise photo made billions of people aware of Earth's cosmic boundaries. Anyone schooled in permaculture would want to explore further, to better understand the nature of the light, tides, meteors, charged particles and other flows across that blue boundary.

So much for fears of "going native." Now what about peace? It is almost a cliché in the tech world (whose university laboratories have grown rich on defense funding) to point out the debt that modern progress owes to the engines of war. This is nothing new. The adventurous have long feared the peaceful, contented apathy of the narcoleptic, lotus-eating Lotophagi described in Homer's *Odyssey*. By eschewing conquest, permaculture and gaiomes would seem to veer in that direction. But as practitioners of ecological living can tell you, any savings in labor comes at the cost of attention and study. Sustainability requires such great care that humans have seldom mastered it for long. Beyond Earth, it will require more attention still. Gaiomians may know peace and contentment, but to survive and move about the cosmos, they must never cease to question, observe and act: the antithesis of apathy.

Like the Tikopians, the traditional Ladakh, the Amish and the Dukhobor, gaiomians would live close to the land in tight-knit villages. They would deliberate carefully in choosing among technologies to meet their needs. Yet, to be effective in space, they would need to know telerobotics, orbital dynamics, computer-aided machining and other advanced technologies—skills not found among traditional agrarians.

No such people exist today, but surely present trends will produce them. Technologists love to use their hands: many enjoy crafts such as home improvement, gardening, knitting and cooking. Conversely, I've met many in science and technology who credited their success to the ingenuity and industry they learned during a rural childhood. As the importance of producing food locally with little expenditure of energy and fossil water supplies comes to the fore, the status of small farmers and gardeners will rise, just as it has in Cuba. Modern communications infrastructure allows permacultural innovations to spread rapidly throughout the world. Intentional communities are an increasingly frequent response to the rootlessness of modern life. Thus it is likely that science villages will emerge that are astute about both technology and ecology.

Why would a thriving eco-community ever leave the rich lands it married on Earth? Why abandon Gaia's vast reservoirs of air, water, minerals and biodiversity? I don't know, except to note that some always have abandoned their homeland. Even native Americans whose ancestors have lived harmoniously with the land for thousands of years are descended from wandering Asian tribes.

The Greek word for wanderer is "planet," and gaiomes would be living worlds in their own right. Combining architecture, technology, gardening and art, they would prove irresistible not only to the individual dreamer, but to the occasional established village. The wanderer trait lives in many of Gaia's species, from sea-birds to a few of our earliest human ancestors. It's part of who we are, and, if Gaia is alive, who she is.

To illustrate these points, let's take a brief fictional journey.

Astronesia

The residents of Cattail Village trace their roots back to a suburban commuter hamlet of the same name on Old Earth. As once-prosperous tech industries left their region, many of the locals found themselves out of a job. Houses went into foreclosure and remained vacant, but the Cattails, fortunately, had knit together as an ecology-minded community. Displaced families crowded into converted garages, attics and basements. Together with several nearby developments, the Village established a local currency and diversified their zoning for mixed commercial/residential use. Gardeners set about rehabilitating soils long damaged by lawn chemicals and began growing food and lumber for the local market. As energy prices rose, the Catttails sold most of their vehicles and established a car sharing co-op on the edge of the village.

Next, they dug up and planted large sections of street, leaving small paths (inspired by Dutch *woonerfs*) for pedestrians, bicycles, electric runabouts and the occasional car. Houses began sprouting greenhouse additions, which allowed them to grow food throughout the year and cut heating bills in half. As refrigerators and furnaces expired, they were replaced with integrated geothermal heat exchangers, which, along with better home insulation, cut their power demand by a factor of ten or more. Finally, the village installed solar photovoltaics and wind generators. Soon they were not only energy independent, they actually exported electrical power for profit much of the year.

Life was good in Cattail, which by now was nearly self-sufficient. The air was clean; water and food were healthy and local. Gardens, climbing trees and natural swimming ponds abounded. With no commutes, the young and old had time to work and play together. The arts flourished. Wildlife was returning in a widening outer zone and along a wavy creek that ran though the village. Yet an ugly incident between several village youth and a lone traveler woke the residents up to two unpleasant facts: first, nearby cities were suffering, and second, they, the villagers, were becoming provincial.

Provincialism and restless youth proved fairly straight-forward to handle. The cure in both cases was frequent exchange programs with families from big cities and foreign countries. As a result, Cattail residents learned twenty different languages and became known for their skills in salsa dancing, pine needle basketry and gamelon. But the plight of nearby cities, still struggling in the extractive economy, proved more difficult.

Though many cities around the world were successfully making the transition to the solar economy, the nearest weren't. Their residents were dumping wastes nearby and buying meat from gigantic confined feeding operations that polluted local water supplies. The few who fled into the countryside tended to buy large plots of vacant farmland, mow them into lawns and build huge, poorly-made homes. This put enormous stress on the remaining wild spaces. But what could be done? The topic cropped up often in village meetings.

Several of the villagers had pooled their savings early on to buy a hundredth share in a passenger spacecraft. This earned them substantial income, some of which they had reinvested over the years to increase their share to a tenth, or some 150 tickets to orbit per year. Every now and then, one of the gaiome guilds would pester them for gardening tips, but they were too busy with matters at home to get too involved.

Years later, when three teens learned of their parents's rocket venture, they began to frequent the wikifora on the Interplanetary Internet and read up on space travel. There they learned that Mirasol, a new Astronesian gaiome, had come to L_1 and was looking for three stable villages from Earth to come settle it. Their enthusiasm for space travel ignited on the spot, though it took awhile to infuse the rest of the village. People began to talk about moving out to a gaiome and selling their village entirely to city evacuees. Passion ran strong both for and against the move, so the rocket owners granted the most vocal on both sides passage to the gaiome for a one-month fact-finding trip.

The fact-finders returned that spring with pictures and stories that excited the villagers. In the end some two hundred people, half of the population, chose to apply for settlement. By mid-summer, eight Astronesian observers (including three children) came to live and work in the village. Dozens more visited for a few days at a time throughout the year.

What were they looking for? Beyond the minimum requirement of unblemished environmental ethics and a good fit to Mirasol's ecology, the Astronese looked for peace, cooperation, adaptability, groundedness: space is no place for the desperate. It also helped that the villagers had many useful skills such as midwifery, dentistry, electronics and plumbing.

The following spring, all visiting Astronesians enthusiastically vouched for Cattail. Several stayed on to help coordinate the move.

By now, word had spread among cityfolk, who got on waiting lists for homes in the village and drove in on weekends for apprentice work (this also provided them with a welcome supply of fresh produce).

The migration commenced that winter. Although the fixed cost of the rocket flights had long been paid, the operating expenses for launching an extended family of six still came to about 80% as much as they would recoup from the sale of their homes. Half of the remaining funds from the sale went into the village trust and half went into medical screening and provisioning. Over a span of three months, thirty rocket launches carried the first wave of 75 villagers and their few belongings to a LEO station. From there, the Pull Economy took over. Astronesian PIT tugs took them out to L_1, where the children crowded about windows trying to pick out the eight gaiomes in their halo orbits from among the stars. Their own gaiome was a two-kilometer sphere, with ten villages of roughly two hundred native Astronesians already established. The gaiome would remain in halo orbit for the next two years as the rest from their village (plus settlers from sixteen other villages) sold their terrestrial homes and joined them in orbit.

In Mirasol Gaiome, the Cattails were entrusted with a small swath near one of the founder villages. The land was only six years old, yet its lush and varied grass and brush were already giving way to fruit and nut guilds in the midlands, where their gardens and homes would cluster. Their dwellings came with an adequate forage of greens, fruit, nuts, beans and grain, along with access codes to their House Journal. Here they would detail their experience building their homes, planting and growing their gardens and landscapes, the presence or absence of wildlife, and other indicators of the health of each zone of their home. In keeping with custom, nearly all households would diligently keep their journals current and publish them for comment throughout Astronesia.

Lacking any threat of adverse weather or earthquakes, they could build lightly. The first villagers arrived to find the roofs up on enough structures to shelter them from the light pre-dawn rains. Over the coming weeks, the locals helped them get acquainted with the chief structural material—bamboo—from which they built furniture, implements, clothing and shelters just thick enough to dampen noise. The bamboo grew in the village and could replace itself after cutting every sixty days. Not only was its wood strong and fibers soft, but it could be composted gracefully when its cycle of human use ended. Although hardwoods growing throughout Astronesia would be used in later dwellings, the bamboo style would dominate their architecture for centuries.

With an air reservoir nearly a thousand times less per person than on Earth, fire prevention was important. The villagers had to get used to cooking in shared kitchens rather than over home stoves. Their few electric appliances were sealed and plugged into inductive outlets that could not spark. The thick, multipurpose electrical/communication cable was built into foundation piers for each home site.

After the final launch, the Rocket Trust operated for eight more months to repay Cattail's modest residual debt and help purchase some nearby crop-land for reforestation. Then it transferred to a loose confederation of villages planning to move to the gaiomes. Within a decade, the Trust would come to own some seventy rockets, transporting not only settlers, but also Gaiomians wishing to visit Earth. Early on, some visitors would bring a few kilograms of asteroidal platinum for spending money and to help defray the cost of a return trip, but over time, the Trust took care of those expenses, too. Within a century, the rocket fleet would grow well into the thousands, carrying millions of settlers to the gaiomes every year.

Now that all were aboard (except for those families—two Astronesian and one from Cattail—who decided to return to Earth), the gaiome set out for the asteroid belt. The voyage would take three years.

Finally they reached Astronesia, a small nation that included a large toroidal gaiome and five smaller spheres from two to four kilometers in diameter. Together, they orbited within two hundred kilometers of a half-mined Aten asteroid, itself four kilometers across. The population at the time was about 200,000, with 3/4 of the people living in the two largest gaiomes.

By the time of their arrival, the new villages of Mirasol Gaiome had built their homes, established perennial gardens and developed a cadre of trained apprentices in telerobotics, astrogation and space manufacturing. As the shifting constellation of sister gaiomes gradually hove into view, the settlers found the time to enjoy and explore their new green world, with its abundance of flora and fauna, ponds, waterfalls, rapids, hidden caves and the occasional window to the outside. The modest attention their hundreds of garden species required left them with lots of leisure time. The arts flourished: music, basketry, textiles and low-gravity dance. The sciences took root, too, with participation by practically the entire populace. Everything from family garden journals to manufacturing technique to metallurgy to meteoroid detection and mitigation studies appeared as data, theory, analysis and debate among the Astronesian wikifora.

Each village made its own food, shelter and clothing and recycled its own wastes. Each had its own arts, festivals and rites of passage in addition to the many customs and observances that were common throughout Astronesia. Each also specialized to some extent. One village had several medical specialists. Another had five talented musical ensembles. Yet another excelled in science. Because their numbers were small, they could exchange services freely, receiving performers and professionals lavishly as guests. When any question of equity came up, just about any third party felt empowered to mediate a fair exchange (a custom borrowed from traditional Ladakh).

Adults and older children briefly but pointedly shunned any child who showed aggression or anger, a tradition common among peaceful societies such as Tahiti. Because of this and the near constant affection shown to children by young and old, male and female, adults never fought and seldom argued (except, per cherished tradition, in the heat of scientific debate).

Children generally had all the skills needed to survive by the time they turned seven: they knew how to find or grow food, swim, avoid troublesome plants and animals and handle their own hygiene with ecological responsibility. They also could find their way around the wikifora, which were safe owing to a complete absence of advertising.

With no streets or harmful vehicles, children were welcome everywhere. Even teleroboticists took turns holding babies as they performed maintenance on the hull. Children, recognizing themselves as full members of their world, hardly ever whined, but watched, listened, asked, mimicked, looked out for each other and learned as is their natural tendency. Almost without effort, some soaked up music, others math or languages, still others weaving or architecture. Though some became prodigies, most children gained adequate skill at so many things that they hardly ever felt left out. They especially knew to seek out the old, who had more to tell of the worlds, and perhaps a bit more patience about telling it, too. Not surprisingly, Astronesia had no schools, jails or malls.

Mirasol's new villagers were eager to travel among Astronesia's other gaiomes. Travel was simplicity itself. Travel pods mounted to docking ports at their equators simply dropped away at the right moment. Momentum then carried them to the equatorial docking ports of the neighboring gaiomes within twenty to forty minutes.

Getting permission to visit was not a problem: all Astronesian gaiomes were based on the same basic ecology, so there was no fear of adverse species introductions. Earth-side screening and treatment kept communicable and infectious diseases in check. However, to stay more than a day, visitors did need to make prior arrangements for reciprocal exchanges with families elsewhere. There was no such thing as uninvited tourism in Astronesia. But given the wealth of arts and wikifora contacts, most villagers had many invitations waiting upon arrival. Trade, as such, was frowned upon. The Astronese preferred to think of life as bounteous and giving, not reciprocal and demanding. There was no such thing as patent, trademark or copyright, and artists and scientists delighted in borrowing from each other. Gifts and celebrations abounded.

Astronesians, of course, had access to contraception, but they seldom had to be very strict with it to keep their population in check. Their near-vegan diet delayed the onset of puberty until the very late teens, and fertility remained low during the two years that a woman typically breast-fed her infant. Moreover, gender equality and a wealth of activities, choices and social commitments each contributed to lower birth rates. The land was not divided from one generation to the next; instead, people flowed by marriage or adoption among various households. When a family line got too thin, village councils (whose members were drawn periodically by lot) would confer about how best to fill its vacant home sites. As with many permacultural societies, the population within each gaiome fluctuated well within the bounds of its carrying capacity.

As the generations flew by, Astronesia mined more of the asteroid to build another thirty-five gaiomes. All were small (two kilometers) in order to make the best use of the asteroid's scarcest resource: nitrogen. The villages that settled the new gaiomes came primarily from Earth. But all quickly became Astronese. The vibrant nation now numbered nearly two million people.

Eventually, though, Earthside villages from other biomes and cultural traditions began to complain that Astronesia was monopolizing its asteroid. Never mind that thousands of other NEOs remained unexplored; theirs had a proven composition and enough material to build at least twenty more gaiomes. After much internal discussion, Astronesia agreed. As a token of good will, the Astronese built a full set of manufacturing facilities, solar arrays, tugs and a cycling gaiome to ferry the next builders between Earth and the asteroid. Then, with a stock of detached arrays, volatiles, metal and glass, they shifted to an orbit that would not directly encounter the asteroid again.

Unfortunately, Rabbit Paw, the second of the new gaiomes at the asteroid, was settled by a town that had not fully overcome its original competitive, hostile culture. Hierarchy and regimentation impeded information flow and decision-making, leading to ecological neglect and, inevitably, crisis. Genetically modified organisms and quick-fix species introductions from Earth only made matters worse, creating, in effect, new pests and diseases. Astronesians flew in on fast tugs to help stabilize the ecology; prior to returning, they had to leave their clothing behind at the docking port and take decontamination baths. During their trip home, they were quarantined and carefully screened for invasive spores, seeds and microbes. It would be centuries before science could replace these drastic biohazard containment techniques with less intrusive measures.

Eventually, some members of the failing gaiome set off for Astronesia in fast tugs, seeking harbor. This was granted (after taking biological precautions) but the new guests proved quarrelsome. Tired of the frequent shunning, most chose to return to Earth. Half a century later, their descendants, proud, armed and angry, set out in a fleet of captured fast tugs to conquer Astronesia.

There was no hiding their intent: lines of sight are long in space, and longer still among the wikifora (the old joke "don't everybody talk at once" kept popping up in discussions of the situation). Before the fleet had crossed the Van Allen Belts, the Astronese began pleading with the pirates: surely it was madness to force their way into a gaiome uninvited. That would risk breaching the hull and losing the irreplaceable atmosphere. To which the pirates replied that they would be happy to "pop" the gaiomes one at a time with a huge drill until the rest, in terror and sorrow, chose to surrender. In reality, the pirates had brought much more sophisticated weapons: computer and human viruses, swarms of parasitic drones to rebuild and reprogram the Astronese telerobots, gamma ray lasers, depleted uranium seeker mines and dozens of other horrors.

There were, by now, over ten thousand gaiomes in space, three hundred of them near L_1 and L_2. The pirates knew to keep their distance, though, exiting on an expensive vector well away from any possible "choke points." Their high-ΔV trajectory was timed to take them to Astronesia in 50 days. Only five gaiomes lay close enough to their path to intercept them with fast tugs bearing gigantic nets of metal feedstock. These the pirates easily dodged, feigning surrender at key moments to delay and thereby widen the error ellipses calculated for their intercept trajectories.

The gaiome defense involved no generals and no secrecy. The pirates laughed as hysterical chatter and pitiful, repetitive pleas for mercy streamed in from throughout the solar system.

The chorus grew however, everyone broadcasting at once until nine thousand gaiomes had diverted all available power to a sharp, loud plea on a single band. The pirates found themselves awash in radio power seven times as intense as sunlight. PITs could take it, but other equipment couldn't. Antennae, connectors and computers died. The aggressor fleet fell silent and overshot its mark. Two months later, teleoperated rescue ships found them. After lengthy quarantine, examinations and inoculations, the raiders were separated and sent to live in computerless villages scattered among several gaiome nations, none of which spoke their language.

Over the next twelve centuries, seventy more pirate fleets of varying size and subtlety set out on voyages of conquest. Fifty-two fleets fell to gaiomic defenses en route to their targets. Among the eighteen microwave-hardened fleets that did manage to conquer gaiomes, four went native, adapting peaceful ways within a generation, four died of influenza (to which the natives had developed partial immunity) and ten used bioweapons of their own to exterminate the native populations. However, none of these "successful" conquerors lasted more than a decade. Seven suffered ecological collapse; the other three soon fell to civil warfare and breached their hulls.

As hundreds of thousands of diverse villages and small nations prospered self-sufficiently on Earth and among the gaiomes, would-be pirates had a harder time provisioning for war. The message became clear: life demands cooperation much more than it does competition. Conquest and consumptive growth lead to failure. In rare cases when a gaiomic or Earth-side village or nation failed, it became honorable and expected for its survivors to renounce their ways, disperse as individuals among the thousands of other settlements or gaiomes, and leave the physical salvage to others.

Still, there were problems. Nitrogen was relatively scarce, even among the C-type asteroids near Earth. This limited the air volume of a gaiome. The torus design reduced the amount of air needed per unit area, but most people preferred to live in spheres. For efficiency's sake, that meant they had to build small. As it turned out, though, the smaller gaiomes (with populations between 1,000 and 10,000 living among five to fifty villages) were also the most socially stable. At this scale, gift and barter economies worked efficiently, inequalities could not get very large, and individual gaiomians could readily perceive the ecological and social consequences of their actions.

Over time, the Inner Zone gaiomes learned to build solar kites and use them to move to the wet, dim Outer Zone. Often they used PIT tugs to move the remaining parts of the asteroid to a safe orbit or take it with them, reducing the threat of impact with Earth. As they got farther from the Sun, their kites grew to compensate for the dwindling illumination. Among the outer Main Belt asteroids and Jovian Trojans, they now could find nitrogen to complement their stocks of other volatiles and metals, allowing them to build new gaiomes with much less waste. But at these distances, they could seldom recruit settlers from Earth. Nearly all were native to their little worlds and few ever visited Earth.

Time and microbes did their work. As generations flew by, gaiomic soils grew to rival Earth's in complexity, subtlety and durability. Home is where the ground lives, and five thousand years after the dawn of space travel, nineteen billion people—95% of humanity—lived in space.

The Far Future

As Gaia and perhaps a few million of her daughters turn the next page in the Book of Earth, questions of identity and evolution move to the fore. How would the tiny ecologies change and evolve in their new environments? How far can they go in the universe? How would they relate to any alien life forms they might encounter? Can gaiomes themselves be considered a new form of life? What would this mean for their human inhabitants? What would it mean for Gaia?

While any answers we might conjure are of course tentative, they do provide a substantial alternative to Khardachev's ravenous cosmic empires.

Free Growth

Once built, gaiomes would need only sunlight to live free of the nightmares of extractive economies. As autotrophs, their survival would not depend on consumption, growth or reproduction.

Still, given the opportunity, life tends to grow. Whether motivated by art, religion or some other creative drive, at least some gaiomes would occasionally use the minor planets they encounter to build copies of themselves, or even try out new designs.

How far could this go? How full can the solar system get? Its ultimate carrying capacity derives from the amount of solar power available. As Tsiolkovsky has noted, the Sun produces 2.2 billion times as much light as reaches Earth—enough to support 10^{17} two-kilometer gaiomes.

There are other limiting factors, however. A 2-km gaiome would mass 120 million tonnes. The main-belt asteroids contain 4,800,000 trillion tonnes of material,[221] and the Jovian Trojans may double that. Altogether, that's enough to make up to 80 billion gaiomes—an impressive number, but still 1 million times less than the carrying capacity of available sunlight.

While 80 billion gaiomes may seem like a lot to crowd between the orbits of Earth and Jupiter, their average separation if they filled this spherical volume works out to 280,000 km—a little less than the distance between the Earth and its Moon. Collisions would be rare: with no traffic planning whatsoever, the average gaiome with its 25-km solar arrays would need to dodge a neighbor once every 20,000 years or so.[222] Still, with 4.4 million near-misses each year, gaiomes might do better to observe some orbital "rules of the road." For example, gaiomes at the same mean distance from the Sun could work their way into circular orbits in a shared plane with a pre-determined inclination and orientation.

Groups of gaiomes sharing language, culture, climate and species mix might maintain orbits much closer to each other than the average separation. This would reduce the number of gaiomes they would need to track and shorten travel times between gaiomes within each "nation." Journeys by travel pods (page 165) between gaiomes a thousand kilometers apart, for example, would take less than three hours. For faster journeys (such as during emergencies), gaiome-mounted PITs could push a metal focal plate on each pod with very little loss of propellant.

What would life be like in a solar economy of 80 billion gaiomes? With human populations of 9,000 apiece, that comes to 710 trillion people, or 110,000 for every person alive today. Discoveries, songs and stories would emerge thousands of times more often than they do today, crossing the solar system via free and open wikifora (and their descendants) in a matter of hours. Yet most people most of the time would probably not tune in, focusing instead on the much more engaging matters of family, village and immediate ecology. Without dependency-creating commercialism to rein them in, millions of autonomous nations would arise, gradually developing their own languages and cultures.

Along with the triumphs of culture, another kind of wealth would grow: the total mass of all living matter. Earth Gaia developed into her present form over the course of well over 10^{43} cellular experiments.[16] As gaiomes increase the size of human populations, they would also increase the total biomass and hence life's overall R&D budget.

Gaiomes would spend this budget wisely, because the majority of them would be much smaller than most Earth biomes. As happens on islands, an individual with an adaptive mutation would comprise a larger fraction of the breeding population, so the new trait would feel less genetic regression to the norm. Thus, tonne for tonne, life would be more able to take advantage of vacant niches and opportunities to co-evolve new forms. When billions of gaiomes dot the solar system, genetic innovations that would have taken a billion years to appear on Earth may occur spontaneously every decade. While any given year may not bring much change in any one place, a few thousand years could give rise to hundreds of times more new biota than the Cambrian radiation.

Niches without precedent on Earth would open up for life among even the earliest gaiomes, the earliest of which would already span an unearthly range of pseudogravity between equator and highlands. As the little worlds face unforeseen circumstances such as isolation, accidents, faulty design or tainted manufacturing feedstock, new environments would appear. Eventually, the gaiomic range of pseudogravity, atmospheres, soils, water chemistry, temperature, pressure, lighting and ionizing radiation would exceed that of Earth by a wide margin, inviting broader and faster speciation among microbes, plants and animals.

Undreamed-of creatures would appear at first perhaps in one gaiome out of a million. Then, with increasing frequency, entirely new kindoms of life would start to emerge on every scale: chimerae beyond any imagined in myth or legend. Among the new species, of course, would come monsters and plagues. But all would co-evolve in human communities and Gaian ecologies. In this natural democracy, voraciousness would tend to self-limit, killing its host (and therefore itself) before it can spread. Ecosystems automatically punish the worst conquerors as their depredations scorch the ground around them.

Humans would not stand apart from this evolutionary pressure. On Earth, mammal species typically last but one or two million years before dying off or evolving into other species. While humans may have taken a little longer than normal to diverge from chimpanzees and bonobos, terrestrial evolution still would most likely transform or replace us within a single volume in the Book of Earth. In gaiomes, we would co-evolve together with the ecosystems that contain us. Because evolution would accelerate enormously among billions of gaiomes, post-human species would appear many hundreds of times sooner than they would on Earth. Some may adapt themselves to high gravity: they'll be short, thick and quick. Others may adapt to low gravity, regaining the prehensile feet of our distant ancestors and evolving longer limbs. Still others may have slowed metabolisms, lengthened adolescence, or adaptations to low-nitrogen ecologies.

Genetic engineering and other forms of human artifice would speed this process further, though there are some limits. For example, some gaiomes, not satisfied with the limited size of their milligravity areas, may wish to slow their rotation and adapt all inhabitants to low pseudogravity. At a hundredth of Earth normal gravity, an athlete really could "leap tall buildings in a single bound." But her body's natural response to low gravity—calcium loss—would weaken her bones to the point of fracture upon take-off or landing. Within a few decades to centuries, it may become trivial to perform gene therapy to nullify this effect. But what about the rest of the gaiome? Adapting each of its thousands of species and their interactions to lower pseudogravity would considerably slow any effort at engineered adaptation. Problems would abound.

Bernal wrote about how coming generations could begin to rebuild themselves mechanically, replacing natural organs, limbs and senses with prostheses until the original body was gone.[61] I don't think this will happen any time soon. Beyond the obvious complexity of our internal linkages, we need to consider our external connections. We evolved and live as part of complex ecosystems. Although we view organisms as individuals, life seldom goes it alone. Even the extremophiles of oceanic thermal vents or antarctic glaciers live in communities spanning multiple kindoms of life. If we struck out on our own and roboticized ourselves, we would lose one of life's most enduring precedents. What is life without ecology?

More likely we'll use cybernetics to achieve much simpler tasks such as high-precision, customized production on a small scale. Small, integrated descendants of today's ink-jet printers, computerized lathes and knitting machines would take generic feedstock such as metals, glass, vegetable oils and the like and produce finished microchips, circuitry, mining apparatus, hull sections, etc. Gaiomes would swap machine programs like recipes on the wikifora, improving both their physical autonomy and techno-cultural evolution.

A New Life Form

Would gaiomes qualify as living beings in their own right? Like Gaia, they would have homeostatic metabolisms, an important trait of life. But terrestrial biomes such as forests also have homeostatic qualities, maintaining their own microclimate and bringing thousands of species together to improve cycling ratios. Still, most of us would not consider a forest as a being on a par with a tree or even an amoeba. It's a stretch of current science even to compare a forest to a colonial organism such as a sponge or a slime mold. Maybe this consensus view will change in time. Meanwhile, let's continue examining the evidence.

Clearly gaiomes as described here would reproduce and evolve adaptations (such as solar kites) to new environments. These are suspiciously life-like activities. As autotrophs, they would convert energy but not matter—except internally or during reproduction. Via wikifora and direct visits, they would exchange information, including genes and tooling instructions, much like bacteria have done since long before the invention of sex.

It appears, then, that gaiomes probably would meet anyone's definition of true life forms in their own right. But where does that leave *us*? As components of a larger organism, would we end up as mere cogs, performing biological services for our gaiomes like mindless little organelles? What about free will? What about consciousness? What about creativity? Would we end up trading it all in for a drudgerous role in something not just bigger, but incomprehensibly alien and indifferent to our whims?

Cole and Cox called this new being Macro-Life and assumed (as had Isaac Asimov eight years earlier) that living within it would mean more regimentation and less freedom.[166] George Zebrowsky imagined that minds would merge and lose their individuality.[223] Would this necessarily come to pass?

Perhaps we can learn from the prokaryota, who faced a similar situation two and a half billion years ago. When the Oxygen Catastrophe drove all of them into anaerobic niches, some languished in bogs for two billion years until finally finding jobs fixing nitrogen for plants, helping with animal digestion and so forth. Meanwhile, other prokaryotes took refuge inside a new form of life, the soft-walled Eukaryotes, themselves evolved from prokaryotes. The refugee organisms eventually found or more likely invented work for themselves as organelles such as chloroplasts and mitochondria. This arrangement worked out well for everyone. Today, a substantial portion of your body weight is mitochondria, and you could not move a muscle without them. Prokaryota were always very diverse, yet they have retained their essential features: a hard cell wall and no nucleus. They never did learn to breathe oxygen.

By evolving new roles for themselves alongside or inside other organisms, prokaryota found refuge and ways to reach enormous new potentials. They thrived, evolved and remained creative and adaptable for billions of years. They never became slaves. They never lost their DNA. If we follow their example in gaiomes, humanity has nothing to fear.

Human intelligence and creativity are secure in space because nothing on Earth can breathe vacuum. Only we can build and maintain secure vessels for air, gravity, water, soil and light. Only we can adapt them to the cold, dim Outer Zone. With machines alone, humans can survive in space, but only for a short while and in a brittle sort of way. To live there long-term, we need Gaia's genius for life support. Earth ecologies, in turn, need our cleverness and artistry to survive in space. Together, like merging prokaryota, we can retain our best features and become something more.

Dispersal

Slowly, using kite mirrors for transportation and power, gaiomes would disperse throughout the solar system, out among the main belt asteroids, the Jovian Trojans, past Saturn and Neptune (with, perhaps, their own undiscovered Trojans). Gaiomes with 1,000 km solar kites should be able to travel and survive all the way out to 300 AU: well beyond the Kuiper Belt.

The potential is enormous: together, the Kuiper Belt objects have perhaps as much as 100 times as much mass as the Main Belt asteroids—enough to make four trillion gaiomes. [14,160] By filling the spherical volume out to 300 AU, the average distance between them would expand to several million kilometers. Radio signals would take several days to work their way from one end of the inhabited solar system to the other.

In this ice-rich, light-starved realm, gaiomes would need large solar mirrors to survive. Some may stretch their range (as life often does), ending up too far from the Sun for their mirrors to provide adequate warmth. They may develop ever larger mirrors to compensate, or life itself might adapt, developing cold-tolerant guilds and slower metabolisms.

Thus prepared for deeper space, well-adapted gaiomes would continue to disperse outward. Perhaps, as Sagan and Dyson have suggested, travelers would follow long-period comets into the Oort Cloud and beyond. [224] At 50,000 AU, the Sun would still be the brightest star in the sky, but not by much. With 2.5 billion times less sunlight available than near Earth, even the smallest gaiomes would need mirrors with diameters larger than Jupiter to maintain Earth-like climates. But life, by now adapted to cold, milligravity and possibly vacuum, would be ready for a lower-power existence. With ambient temperatures hovering just a few degrees above absolute zero, ice becomes steely hard, perfect for building transparent low-gravity, low-temperature structures. With metabolisms slowed several hundredfold, gaiomes might require mirrors only a few hundred kilometers across. A comet's natural gravity might have the same subjective quickness to its slow-moving, long-lived inhabitants as Earth's does to us now. These worlds need not be hollow, nor spin for gravity. Residents could dwell on their surfaces under stars that seem hundreds of times closer than they do to us.

The total mass of comets in the Oort Cloud is unknown, but estimates generally place it around 1–3 Earth masses,[168] enough for a hundred trillion (10^{14}) gaiomes. Paradoxically, this vast population would intersect even fewer of the Sun's rays than its Inner Zone counterpart due to its great distance from the Sun. Gaiomes would be 2 AU apart on average, and radio messages would take nearly two years to travel from one end of the cloud to the other. Together, all of the gaiomes throughout the solar system would obscure less than a thousandth of the Sun's light—a comfortingly thin canopy compared to Earth's dark jungle floors. There would be plenty of Sunlight for all.

Oort Cloud comets take up to ten million years, or about two volumes in the Book of Earth, to orbit the Sun. Over the course of several cometary orbits, the Sun gets close enough to nearby stars for them to impinge on the cloud, stirring it up and stripping away a few comets. These, and any gaiomes that dwell among them, would become galactic wanderers.

Long before their random stirrings turn distant gaiomes into star-farers, others closer to the Sun would, no doubt, embark on much faster, much more intentional journeys to the stars.

Many hands make light work. Never was this more true than of star flight. Suppose that at some future date a million Inner Zone gaiomes get behind the idea of star travel. By lot, they choose a volunteer gaiome to send to a specific star. Next, they employ a technique that Robert Forward has called "Star Wisp:" the volunteer uses a thousand-kilometer woven mesh sail to ride a powerful radio beam on the same principle as a solar kite.[225] Because the spacing of its weave is much less than the radio wavelength, it reflects the beam like a solid mirror. Because its mesh is mostly empty space, its mass is much lower than a solar sail of the same size. A focused microwave beam from a single gaiome might be only a hundredth as intense as the sunlight falling on it and its solar arrays, but the combined beams from a million gaiomes could push the sail ten thousand times harder than sunlight. The volunteer accelerates 30 km/s per day. Within five years, it's traveling at ten percent of the speed of light—fast enough to reach the nearest star in 44 years (slowing down is left as an exercise for the reader's distant descendants).

If the volunteer comes from the Outer Zone, slower metabolisms might shrink the perceived trip time to a small fraction of a life span. Gaiomes in transit would power their ecologies either with star mirrors or a trickle charge from their mesh sails. Interior life would proceed uninterrupted. Upon arrival at the new star system, the volunteer gaiome would have no pressing need to prospect or reproduce, though some, of course, would choose to do so.

Only a small minority of gaiomes could take the fast road to the stars. But once in orbit around Red Dwarfs, blue supergiants, protostars and other exotica, they would continue to evolve, adapt and (yes) grow in unexpected ways.

Extraterrestrials

It has become fashionable again to speculate about whether there is life beyond Earth. So far, no one has found any, so the most honest answer is, "I don't know." People hate to hear this, and opinions run strong either way. But without evidence, "I don't know" is all we have.

Sooner or later, gaiomes would encounter alien life forms—if only because the inhabitants of some would eventually evolve beyond recognition. In our extractive economy, aliens would pose serious problems because, once discovered, our way of life puts us in competition with them for dwindling resources.

Of course, we seldom see ourselves as aggressors. I remember a lively tea-time debate on this topic among the Princeton Astrophysics faculty. It was agreed that because we are probably much younger than any civilization we detect by radio search, our radio signals won't have traveled as far as theirs. Thus we would most likely become aware of alien civilizations centuries before they become aware of us. Department Chairman Jerry Ostriker then asked: what if they turn out to be technically advanced, yet voracious and predatory? What could we do in the century or two that might pass before they discovered and set out to conquer us? People chimed in with all sorts of proposals for interstellar weaponry, such as shooting magnetic monopoles (a fictional subatomic particle) into their sun; somehow this was supposed to cause it to supernova. Others suggested just amplifying our television sitcoms to export our destructive consumer culture.

Indeed, some gaiomes may rediscover extractive economies upon arriving at new star systems. They may become galactic marauders for a time. In so doing, however, they would be hard-pressed to retain the advantages of autotrophism and survive on the cosmic scale. They would be swimming against a very stern evolutionary tide.

It seems likely that gaiomes would give non-intelligent alien life wide berth. As for intelligent life, gaiomes could try to interact by swapping those things that do not diminish when shared: stories, science, music, drama, images, recipes. It's certainly worth a try. Who knows if it's possible to communicate with species that followed completely different evolutionary paths. Will extraterrestrials be as inscrutable to each other as algae are to zebras?

Because of their self-sufficiency, gaiomes would not require parity when trading information. Even if your gaiome transmits stories and recipes a hundred times more valuable and voluminous than those it receives, you would most likely find the exchange beneficial, if only perhaps to your self-esteem. Information sharing would not alter your competitive position as it would today on Earth because gaiomes won't depend on trade for survival.

On the other hand, if another gaiome seems to have much to teach you, it might be prudent to receive its transmission with gratitude, but treat the contents with care. When religious farm communities such as the Amish encounter a promising new technology, they deliberate over whether it contributes or detracts from the way they want to live, work and raise their young. As a result, many still ride horse buggies and get by largely without electricity; yet they do put high-tech reflectors on their carriages to protect themselves from motorists. Similarly, it would be prudent in a gaiomic exchange to ask whether each new technology, art form or cultural practice would help or hinder your way of life.

One form of information in particular would require careful handling: genes and their expression in living beings. Gaiomes should consider any exchange of plants and animals with care. New species introductions have left an indelible mark on every continent of Earth over the past few centuries. With their smaller physical and biological reservoirs, gaiomes, like islands, would have a harder time absorbing an adverse introduction.

Gaian Evolution

As autotrophic, Earth-derived worlds in their own right, gaiomes would seem to be legitimate children of Gaia. Their name even implies this. But if Gaia is alive, perhaps she already has found other ways to reproduce.[226] Perhaps she even evolved from some prior life form.

This would seem hard to argue. Lovelock and others have described Gaia as a super-organism shaped by the coevolution of all species together with their climate and geology. One could say that Gaia is her own ancestor: creating, surviving and being shaped by catastrophes—many, such as the Oxygen Catastrophe 2.5 billion years ago, of her own making. To be evolving in the sense that her component species are, she would need to inherit traits shaped by cosmic selection pressures. She would need ancestors beyond Earth. Yet none of the thousands of moons, planets, meteorites, asteroids and comets studied to date show any signs of past or present life. At the very least, life is rare in the solar system.

Nevertheless, meteorites have been found on Earth with isotopic compositions unique to Moon and Mars rocks. This raises the possibility of *exogenesis*: life spreading from planet to planet by natural space travel.

Recent years have seen a lot of work on this topic. Computer models show how asteroids impacting a planet could act like hypersonic bulldozers, tearing tens of thousands of rocks completely free of a planet. Martian meteorites so liberated might take decades to millions of years to reach us, depending on where they enter a resonant dance with Jupiter's gravitation. Further calculations show that asteroid impacts on Earth should also launch rocks into space, seeding the other planets with small boulders and chunks of topsoil over time. With statistical inevitability, some of these meteoroids would find their way to Jupiter and experience the same sort of gravitational swing-by that sent the Mariner and Voyager probes out of the solar system. Thus, every few years, a clump of rock or soil from Earth would find itself hurtling star-ward. [227]

But it's a rough journey. The several hundred thousand gees involved in being blasted off Earth's surface would pulp any animal or plant. However, experiments with bacteria-tipped bullets have confirmed that some microbes can survive these accelerations. [228] Although the surfaces of ejected boulders would also experience searing heat, the interiors of many would never reach the $100\,°C$ needed to sterilize them. Similarly, while the raw ultraviolet light from the Sun would damage exposed cells within 10 seconds, it does not penetrate more than the width of a single cell. The vacuum and cold would cause freezing and drying, but common bacteria such as *Bacillus subtilis* have long-since evolved a dormant, spore stage that can survive drying for decades and freezing essentially indefinitely. [229]

Time, though, is not kind to bacterial spores. Even frozen in a protective layer of ice, their DNA slowly degrades due to chemical reactions with oxygen and the natural decay of radioactive elements within and around them. Solar charged particles and cosmic rays have a similar effect. Combined, these destructive processes would kill about half the spores of *B. subtilis* every 33,000 years, and *Deinococcus radiodurans*, the most radiation-resistant bacterium, every 140,000 years. [229]

While this is plenty of time for Earth to infect Mars with living rocks (and visa versa, if Mars ever had life), it's much too short for interstellar travel. Simulations suggest that Earth rocks would careen around the solar system for some 20–80 million years before a chance encounter with Jupiter or Saturn could eject them. [227] By then, they would be completely sterile.

But the universe is messy and life is clever. Bill Napier, an Emeritus Researcher at Armagh Observatory in Northern Ireland, has come up with a tortuous set of routes by which exogenesis still could grant Gaia children, perhaps even a parent. [230] A typical path might go like this:

Every billion years or so, the Sun's orbit takes it through a star-forming region like those seen in Orion. With a mass of a hundred thousand suns, the gravity from this Giant Molecular Cloud (GMC) disturbs the Oort Cloud, raining comets down on the inner solar system before it even leaves the GMC.

Jupiter captures many of the new comets, holding them in the inner solar system until most of their volatiles boil away. The dust left behind increases the ambient mass of zodiacal dust a thousand-fold. The Zodiacal light, normally a faint yellow pre-dawn glow, girds the sky in a band now visible in daylight. The Sun's image reflected in the dust (called *gegenshein*) blots out the Milky Way at night.

Some of the comets strike the Earth, and when they do, they hit at much higher speeds than any asteroid. This, in turn, plows more material than usual into space. Immediately, the zodiacal dust cloud begins to grind away on the dislodged boulders like a celestial sand-blaster. Within a thousand years or so, the boulders disintegrate into tiny grains. This essential step greatly increases the number of potentially life-bearing particles in space, and with it, the odds of seeding a planet.

But first the grains need a faster way to leave the solar system than the wait for a chance Jupiter fly-by. Fortunately, light pressure on a grain depends on its size squared, while the Sun's gravitational pull depends on its size cubed. For small grains (about a micron across), sunlight wins over gravity and pushes them out of the solar system. Within a few tens of thousands of years, they plunge deep into the hearts of the GMCs proto-planetary disks. There, a few of them become entrained in newly-formed comets.

Bacteria caught in the grains find themselves suddenly cocooned in many meters of ice, which protects them from further cosmic ray damage. In this way, passing solar systems could store and exchange bacteria in the GMCs. If all comets formed in such systems as they were being seeded (a *huge* if), the dust from them would carry a new bacterium to Earth every 300,000 years or so—about sixteen times per volume in the Book of Earth.

Napier's theory solves the problem of spreading bacteria and storing them somewhere safe from radiation within only a few thousand years, but it does not explain how the tiny grains would survive the ravages of vacuum for more than a decade or two while leaving the solar system. Nor does his scenario address the chemical and radioactive destruction of the DNA during its long ride in a comet, nor the odds of the bacterium surviving impact and getting established on a new planet. His theory stretches credulity to the breaking point. Still, it's *almost* plausible and it only took him a short while to come up with it. Throughout cosmic time and space, life has had a lot more opportunity to get creative or lucky. Gaia may yet prove to have children.

If the odds of life arising independently turn out to be much lower than those of surviving a seeding event (and passing along the space-faring traits), then Earth Gaia would likely have a parent. Unfortunately, we know neither the odds of origin nor the odds of survival, so we're left wondering.

It would be tempting at this point to say that gaiomes would grant us time and scope to determine whether Gaia has parents or children. However, laboratory experiments and powerful new telescopes will likely resolve the question long before anyone builds a gaiome. Earth, so far as we know, is unique in the universe: a living world startlingly far from equilibrium. Anything like it would stand out clearly from the surrounding deadness.

Even if Gaia has a parent, her evolution has been measured in the billions of years—thousands to millions of times slower than the evolution of its individual species.

With the evolution of human intelligence, Earth Gaia may have found a faster way to evolve. As innumerable gaiomes scatter throughout the galaxy, they would experience a wide variety of selection pressures on much shorter time scales than the Book of Earth. A whole host of environments from the hot inner solar system to deep interstellar space would test every facet of their being from technology to ecology to architecture to political and economic models. The best-adapted to autotrophic life would thrive and likely help others to do the same.

This has profound implications. While Gaia probably does have a metabolism regulated by at least some of the enzymes of the world's biomass,[113] we do not know yet whether she is herself actually alive in an evolutionary sense. Does she reproduce? Does she carry traits from prior generations? Does she experience variation among those traits? Has she survived environmental stresses that progressively select for certain traits? None of these questions has been answered. But for gaiomes, the answer in each and every case would be a definite "yes." Whether or not Gaia "lives," her children, the gaiomes, definitely would in every sense of the word.

Humanity has a very bright future in the cosmos: not as a monolithic civilization, but as part of innumerable ecologies. Humans would not rule the gaiomes, but simply fill niches within them. Our jobs might include exploration, design, hull-building and communication. Nature will always have new challenges for us. As we adapt to local conditions beyond Earth, we'll find our language, lore, science and arts growing usefully richer. As we gradually evolve into many species, we'll come to know the universe better.

Eventually, the Book of Earth will come to a close. Left alone, the Sun will run out of hydrogen fuel and engulf the Earth as a Red Giant. This will happen a thousand volumes from now. Perhaps by then our descendants will have "lifted" the Sun as Criswell foretold, giving themselves a ten-million volume reprieve. Perhaps they will seek refuge among the long-lived Red Dwarfs that outnumber and greatly out-live the Sun and other bright stars. Perhaps they will find new suns among the many born every year in the Galaxy. Perhaps, in their diversity and the immensity of time, Gaia's children will do all of these things. Life will find its own way. That's its genius.

Evolution took innumerable branching paths to produce the rich world we live in today. We and our kin—the animals, plants, fungi, protists, archaea and bacteria—got here together. The know-how to make a living world resides not in labs or books or even one species, but in the sum of all of life's kindoms. Only together can we step into the cosmos. In the long run, who really knows whose genes will take the lead?

Diversity, not power consumption, is the proper measure of cosmic progress. By abandoning Bernal and Kardashev's solipsism, we can discover a future far more varied, lively and meaningful than any we could wrest from the universe. But to open this path to our descendants, we must end our brief foray into conquest and rejoin the family of life.

Chapter 8

Homework

"...I would relate to this land I called home *as if I were responsible for building the culture that the rocks and trees and birds of this place expected of human beings.*"
—Chellis Glendinning[231]

What will it take to live in space? I used to think the answer would come in the form of some clever invention or business plan. But as the preceding chapters have illustrated, our difficulties owe more to our behavior than to any lack of technology or money. When we build spacecraft like missiles rather than passenger vehicles, we end up with expensive, dangerous munitions. When we lob expendable, break-apart boosters and satellites into orbit, beyond reach of Earth's cleansing ecology, of course the debris turns LEO into a shooting gallery. When ground-based solar and wind power are already beginning to compete with coal, why build solar power satellites and extract Helium-3? When we're causing extinctions at a thousand times their natural rate, why expect arks and colonies to survive at all in a realm with no natural capital?

For thousands of years, human civilizations have been overthrowing life's democracy and treating everything they see as a possession. As a result, we have come to live well beyond the capacity of this world to replenish what we consume. There's no conquering our way out of this: until we learn the lessons of ecology, we're stuck on a dying Earth in a dead solar system. To live anywhere in the cosmos, especially here on our home of 3.8 billion years, we must face up to an evolutionary task. The work challenges not only our technology, but our very identity as a species. We need to find niches for ourselves that

239

sustain rather than destroy the biodiversity that makes this a living planet. We need to become native to Earth again.

Millions of people are waking up to this task and its sobering enormity. To prevent the loss of half of Earth's remaining species, to stop global warming, to end the resource wars, and—yes—to have a future beyond Earth, we must immediately and decisively reject the extractive economy and embrace regenerative ecologies. But how?

In America, our first reaction to crisis is war. The news these days is full of wars: on nations, terrorists, drugs, illiteracy and poverty. A few decades ago, President Jimmy Carter even declared conservation to be "the moral equivalent of war." But what use is war when ever-accelerating total competition continuously prepares newer, better-armed crops of tyrants? War is the result of, not the cure for, our underlying patterns of growth and consumption.

Our task, if we want to amount to anything in the cosmos, is not a battle of any kind, but the exact opposite. We need to withdraw completely not only from war itself, but from the tyrannies that spawn it: relentless acquisition, extraction and disposal. Yet at the same time, we should be wary of insularity and Spartan self-denial. Conservation, as Jane Jacobs has pointed out, is not the equivalent of war, but the central ethic of both ecology and economic abundance. The high cycling ratios and tight resource loops that define conservation are hard to achieve through austerity, but life's ability to conserve improves spontaneously as the number and diversity of local, supportive connections increases.

How would would an ever-renewing solar economy work in practice? Economist Herman E. Daly has identified three rules that any economy must follow in order to remain sustainable. It must: (1) produce waste no faster than the environment can absorb it; (2) consume renewable resources no faster than the ecosystem can replenish them; and (3) expend non-renewables no faster than the dividends from their use are re-invested in renewable replacements.[232]

Conventional wisdom holds that shifts of this magnitude can only proceed in a top-down manner, perhaps by writ of the United Nations and the major superpowers. After all, international agreements did manage to stop the manufacture of chlorofluorocarbons once these refrigerants and pressurants were linked to ozone depletion. Now an international consensus document called The Earth Charter (www.earthcharter.org) has emerged to catalog best practices to regenerate culture and ecology. It is, in effect, a Declaration of Interdependence.

But such decrees face stiff resistance. Nearly all political and commercial ambitions embrace growth and conquest, contributing ultimately to extinctions, global warming, warfare, alienation and loss of nature's regenerative services. We have structured competition so adversarily throughout the world that it would be economic suicide for any government or corporate directorship to declare a unilateral end to growth and disposal. Even if such an edict prevailed globally, it would work against locality, that aspect of ecosystems and cultures that keeps nutrients circulating in relatively tight loops, minimizing loss.

Some have argued that commerce should be managed at the level of bioregions. Ecotrust founder Spencer Beebe, for example, has urged residents of the North Pacific Rim from Alaska to Northern California to join an ecological alliance called "Salmon Nation." While the region's ecology from fisheries to forests surely would benefit from coordinated governance at this level, here again, efforts to enact it face an uphill struggle against entrenched short-term interests. Many must come together and reach agreement; this will take time.

There is, however, a place where the problem of consumption is most severe. Happily, it is also where change can begin immediately, without waiting for large political constituencies to meet, debate, align, vote, ratify, fund and phase in. That place of desperate need, immediate action and rapid results is, of course, your home.

If the engines of commerce and politics are the levers that move the world, their fulcrum is your own life. How you choose to make and spend your money has everything to do with how the world is configured today. You probably work at a job that pays many times the world average. The most competitive among the developing nations may hate to admit it, but it is how you define success that they seek to surpass. You are a world leader and you cannot resign.

This chapter distills some lessons from gaiome design into strategies that you can use to leave the extractive economy, embrace the solar economy and so lead the world. You do not have to sign a pledge, join an organization, go to conferences, move to the country or give up all worldly possessions. Far from it. Life is everywhere connected and bounteous. There's still a regenerative ecology all around you. Here's how to rejoin it.

Identify your Impacts

Like it or not, you are a consumer. Nearly everything you buy or rent came from the extractive economy and is destined to cause or become harmful waste. Because you need to consume to live, it may seem like you have no choice in the matter. Not so.

To discover just how many choices you have, start by determining your ecological footprint: the total land area that supports you in your present circumstances. Recall that for gaiomes, this number was 1,000 square meters. To calculate your footprint, go online and visit the Earth Day Footprint Quiz (www.earthday.net/footprint/index.asp). After a few screens of questions, your result will appear as the number of Earths it would take to support every human being if they each consumed as much as you do. If that number is more than one, be sure to click the *Take Action!* link to see what you can do to reduce your footprint. Embarrassingly enough, my own number has soared recently to 4.9 hectares (49,000 square meters), or 2.8 planets if everyone lived like me. Of course I'm working to reduce it.

The Footprint Quiz does not claim to address all of our impacts, only those that could be sustainable. For example, it does not include the effects of discarding persistent toxins such as the heavy metals found in batteries and motor oil, or radioactive wastes from nuclear reactors.

The Quiz also doesn't make clear that if you have children by natural birth rather than adoption, you really should not count them among the total number of people sharing the ecological footprint of your house or car. Population growth is still growth, yet in the quiz, children can dilute the footprint as if they were not a result of your actions. Don't use them to water down your footprint. If you are planning a family, having two or fewer children and having them later in life will go a long way toward alleviating human population pressure.

Even without children, you could never shrink your ecological footprint on Earth to anything like what it would be in a gaiome. In a tiny world, no one would need cars, nor would they have to move food and finished goods over long distances. Without roads or bald deserts and mountains, nearly all of a gaiome's land would be alive and fertile. Without serious storms and other natural disasters, its buildings would need far less insulation and structural materials than Earth-side buildings. Nevertheless, the smaller we can make our footprints on Earth, the easier it will be to live in space. Not only could we build smaller worlds sooner with greater confidence, we would have a much greater chance of surviving long enough to build them.

To become even more aware of the power of your daily choices, go back to page 39 and research the answers to some of the "Where does" questions.[233]

Leave the Extractive Economy

The Footprint Quiz may leave you with disturbing images of the deprivation that seems necessary to share this world equitably. Yet several strategies can reduce your footprint by 50% or more while not only maintaining your standard of living, but substantially improving it. By establishing more beneficial connections with your local ecology and neighbors, you will automatically improve your health, prolong your life and increase your wealth.

The primary strategy concerns your food choices. Clearly, the foods you eat influence both the environment and your health. As it happens, the food choice that is healthiest for the planet is also healthiest for you: local organic vegetables and fruits.

The Vegan Advantage

As we saw on page 148, eating exclusively or primarily vegetables and fruits greatly reduces your ecological footprint in the current food production system. Given that more than half the world's grain goes to feeding cattle, it is clear that humans have several times more land in production than we actually need.

Reducing meat consumption is not just about land. Fresh water is one of the most threatened resources on the planet, and the best way to conserve it is to avoid meat. Beef protein requires six times more water to produce than the same amount of vegetable protein.[234,235] According to a recent UNESCO study, it takes 15,497 liters of water to produce a kilogram of beef[234]—about as much water as many Americans would use to shower for a year.

Your daily food choice is by far the most powerful and peaceful way of changing the status quo. Given that cattle typically are slaughtered after 2 to 4 years and many green vegetables can produce two or more crops per season, your food preferences send a very rapid signal to the economy.

Vegans are vegetarians who choose not to use any meat or dairy products whatsoever. Giving up eggs, milk, fish, pork, poultry beef and even honey may seem an act of deprivation, but the plant world is vast and astonishingly varied.

Just as milk and eggs have spun off specialty cuisines of cheeses and baked goods, so have fruits, nuts, seeds and grains. When you visit a farmer's market or the produce aisle of a large supermarket, you will find hundreds of different kinds of fresh vegetables and fruits—a diversity of quantity, color, texture and taste that puts the meat counter to shame.

Several things change when you become vegan or near-vegan. You can gorge without limit on fresh greens and fruit without ever getting fat. Your digestion improves. So does your mood and energy level. You feel cleaner. Your body naturally settles into its ideal weight. You are far less susceptible to vascular disease, diabetes, cancer and osteoporosis. Your medical expenses are substantially lower.

In *Eat to Live*, Joel Fuhrman, M.D. draws on decades of clinical and epidemiological studies as well as first-hand clinical experience to show how a vegetarian diet low in sugars and starches begun in mid-life can add a healthy decade to your life expectancy.[236] The longevity benefits may be even greater for people who start younger: due to poor diet, most American children already have weakened immune systems (as evidenced by ear infections and allergies) and early-stage atherosclerosis.[237]

If a vegan diet seems too restrictive, you can prepare all-vegetable meals at home most of the time and be somewhat flexible on social, family or ceremonial occasions. Vegetarian and vegan options are showing up more often in restaurants and social gatherings, and just inquiring about them sends a strong message. Many of us have met vegetarians who get combative and sulky when the veggie options are sub-par. This attitude does not help anyone withdraw from a war-like mentality. With integrity, humor and some flexibility, you can lead by example. You never know who among the carnivores might be just a few bites away from going vegetarian themselves. Becoming aware of your preferences may be all the encouragement they need.

But what about nutrition, particularly protein and calcium? In grade school in the USA, many of us watched films sponsored by the meat and dairy industries that explained why you need milk for calcium and meat for protein. The arguments seemed very scientific. Milk has lots of calcium, and early studies on rats found most plants deficient in one or more of the essential amino acids they need to make protein. However, as Fuhrman shows, decades of international epidemiological studies have clearly linked adult dairy consumption with a greater incidence of osteoporosis, not less. As for protein, vegans have long combined greens, grains and legumes to get a complete mix of amino acids. It turns out this was hardly necessary. Rats have different protein needs than humans. When he considered actual human protein requirements, Fuhrman found it nearly impossible to concoct a combination of leafy greens and fruits that was deficient in the essential amino acids.[236] As long as you are getting enough calories in your low-starch vegan diet, you are getting enough protein, even if you are an elite power athlete like Carl Lewis.[238]

If you are interested in vegan nutrition, Dr. Fuhrman's book is a good place to begin. I would also recommend *Becoming Vegan*, by Brenda Davis and Vesanto Melina. If I could only have one vegan cookbook, it would be Jennifer Raymond's *The Peaceful Palate*, which also makes an excellent first cookbook for anyone.

Eating plants rather than meat and dairy does much to heal the Earth and the human body, but our food strategy still has two more components: local and organic production.

Local Produce

The average piece of produce in the USA travels some 2,400 km from farm to supermarket.[239] Given that most of us eat our own weight in food about every six weeks, the fuel used to transport our food could have transported each of us 21,000 km every year, or half-way around the world. For energy reasons alone, it is prudent to eat local produce. As you can verify in the Footprint Quiz, doing so will reduce your footprint by about 4,000 square meters—a modest amount on Earth, perhaps, but not in a gaiome. However, reduced energy consumption is not the main benefit of eating locally.

Mass food production and distribution tends to the inhumane and dangerous because it prevents consumers from seeing where their shrink-wrapped produce comes from. For example, in the fall of 2006, spinach tainted with *E. coli* from a single ranch in California sickened over 200 people in 26 states across the USA, killing at least three.[240,241] None of this would have happened had we habitually consumed our food within a few kilometers of where it was grown. Shorter shelf and travel times reduce the growth and spread of infectious organisms. Local operations are small and diverse, limiting any harms from mis-handling of food. Nosiness counts for something, too: the closer you are to your farmer or producer of any kind, the less they will be tempted to take dangerous short-cuts such as washing tonnes of produce in a giant sink.

A handful of customer opinions, while no match for the PR budgets of major conglomerates, can make or break a small farmer's business

The law in many lands considers farms of less than a thousand hectares to be small, a supersized perspective that perpetuates the myth of "away." How much land does a farmer need to make a living? When customers live nearby, the answer works out to less than a single hectare. For example, in Pasadena, California, the Dervaes family grows over 2.7 tonnes of vegetables a year on a 1/12 hectare suburban lot that includes their five-person house. Farming on their property provides half of the food they eat, plus a substantial income from restaurant sales. You can follow their progress online at www.pathtofreedom. com.

Organic Foods

The final piece of our food strategy is to eat organic produce as much as possible. It is truly bizarre that "conventional agriculture" now refers to the heavy machinery, pesticides and fertilizers that have come to dominate farming only during the past eight words in the Book of Earth. It would be much more accurate to call these practices "extractive agriculture" because they convert irreplaceable fossil fuels into food in a very unfavorable ratio: the machinery and chemicals involved consume seven to ten calories of energy per calorie of food produced.[148] As such, mechanized, chemical-intensive agriculture is patently unsustainable.

In Bloomington, Indiana, farmers tell of buying a pesticide, only to find that nature follows the permaculture principle of stacking functions: as promised, the targeted bugs die, opening the way for other pests to fill in for them. The multiple pest species are doing their jobs as decomposers: converting monoculture crops into something that a much wider range of species can use.

Organic gardening uses nature's tendency toward diversity to improve both yield and soil biomass at the same time. Modern organic methods use the life in healthy soil ecosystems rather than sterile chemicals to grow plants. Even back-yard garden soils, fertilized with harsh industrial byproducts, will die as evidenced by a sharply reduced population of earthworms and other soil fauna. With the introduction of organic methods such as sheet mulches, compost and the like, the soil springs back to life. While any given plant has its pests, methods such as crop rotation, interplanting, polycropping and fallowing that increase plant diversity in the garden prevent the kind of pest population explosions that you see with monocrop farming. Shrewd organic gardeners even grow seed plants and florals to attract selected species of birds and predatory wasps: these further cut down on pest populations.

As their soils and ecologies mature, the yield from organic farms and gardens tends to increase over time. This contrasts sharply with extractive methods, which progressively sterilize the soil, causing a loss of fibrous fungi and plants that hold it against erosion and run-off.

Organic gardening depends on diversity and locality, so it does not lend itself well to mechanization and economies of scale. This has led to criticism that it is expensive and labor-intensive: a boutique cuisine for the rich, perhaps, but hardly a viable way to feed the world. In fact, an accurate energy accounting supports the exact opposite conclusion. Organic gardening produces far more calories of food than the manual labor required to grow and harvest it. For this reason alone, it is sustainable at any income level, while extractive agriculture is not. That the organic approach also restores the biodiversity and stability of the soil, improves nutrient cycling and reduces erosion makes it the only choice for long-term survival.

Where can you find local organic produce? If you can get to a weekly farmer's market, that's a good place to start. Because the market allows you to meet growers face-to-face, you can ask about their use of pesticides, fertilizers and organic methods. Some growers prefer not to take on the expense and documentation requirements of government "Certified Organic" programs, so it's a good idea to ask, even if their booth signage doesn't say organic.

When you do find an organic grower, ask if they also sell to shops and restaurants; those are the ones to visit when you go out! You might also ask whether they have a CSA, which stands for Community Supported Agriculture. In CSA, you pay your grower at the beginning of the season to cover their up-front costs; then you pick up or receive delivery of produce every week. Because stores pay growers only a small fraction of the retail price, CSAs can provide you with top-quality organic produce at very competitive prices. Given that most people only have time to buy from one or two growers, CSA encourages farmers to grow a diversity of crops each season, rather than monocrop. This improves the soil's overall fertility and its efficiency at cycling nutrients.

In summary, eating primarily local organic vegetables and fruit is the simplest and most powerful way to support a bright cosmic future for all species on this planet. If it costs a little more at first, it will repay you many times over in terms of reduced medical expenses and a better environment.

Not Just Food

While you're at it, consider buying locally made, grown or reclaimed clothing, cleaning products and construction materials. The more demand you can shift from the extractive economy to a regenerative ecology, the faster your ecological footprint will shrink. As always, go for quality and durability over quantity. The goal is to avoid owning (and later discarding) too much stuff.

Live Small

Once you are eating and dressing right, the time has come to take a look at your physical dwelling. The trend in the USA until perhaps this year has been toward ever larger houses with ever more numerous amenities such as bathrooms. More and more people are buying houses so big that they can't even get the cleaning done in a day, let alone keep up with the added expenses of heating, cooling and maintenance. The new commuter McMansions that sacrifice quality for sheer size are often shoddily built, which adds to all expense categories. Unfortunately, so many have gone up as to shift most people's perception of what's normal in housing. The time has come for a re-adjustment.

If you are looking for a place to rent or buy, consider the advantages of something smaller. It may cost more per square foot, but your on-going expenses, from lighting to heating to commuting (if it's closer to work), will be less. Of course, the smaller your home, the smaller your footprint.

Over the years, I have lived in small houses, apartments, and even sailboats. The boats were tiny: narrow leisure craft with about 30 square meters total living area including the cockpit and decks. Yet their profusion of built-ins comfortably housed two to four people with 60 days of non-refrigerated provisions and enough books, games and music to keep everyone occupied for a year or more. Through the use of fold-up bunks, drop-down tables and the like, the space would reconfigure as needed.

And so people have lived for centuries in many countries, from French peasant villages to rural Japan. The small, simple reconfigurable home or apartment is both the past and the future of housing. It can have top-quality design, fixtures, flooring, trim and built-ins at a small fraction of what a no-frills McMansion would cost.

You can't fit much stuff inside a small space, and this is a good thing both for sustainability and your savings account. A big house can swallow all your furniture and still look empty. A cluttered, poorly-designed house feels cramped. Whether you end up shopping for more stuff or more space for your stuff, it's all too easy to buy into a cycle of growth, acquisition and waste. In contrast, when a small, well-managed house or apartment becomes a permanent home, it enforces good environmental ethics: you can't just buy something and toss it in the Junk Room until you figure out what to do with it. You end up thinking before you spend, a process that saves scads of money over the years. Your footprint starts out small, stays relatively small, and your fellow species thank you.

For examples of extremely small, high-quality houses, have a look at designer Jay Shafer's work at www.tumbleweedhouses.com. If you are thinking of building a house, consider using natural, renewable materials such as slip-straw and chord wood, and using superinsulation and passive solar design to reduce or eliminate heating and cooling costs. For roofing, consider using metal for rainwater catchment. Although slightly more expensive up front, these choices can greatly reduce your utility costs and ecological footprint.

If you are not presently looking for housing, you can still reduce your footprint and your monthly expenses by taking in boarders or extended family members (if your home is large) and installing energy-saving fixtures and appliances.

Save Energy, Save Money

What if you could earn 300% per year on an investment? In many cities, that's the effect of replacing your most-used light bulbs with compact fluorescents. These have improved so much over the years that most people can't tell the difference between them and incandescent bulbs. They can pay for themselves in as little as four months, giving you the equivalent of a 100% rebate thrice a year thereafter.

You can make these savings work for you by applying them toward further energy upgrades. This has a compounding effect, like interest, but without any taxes to slow you down as you cut your energy bills quickly and permanently. For most homes and apartments, the next logical upgrade is to replace an aging canister water heater with an on-demand or solar heater. An on-demand model can free up about a square meter of floor space, which in coastal real estate markets more than pays for itself on the spot. Then it will, via reduced electric bills, continue to recoup its cost about once every year. Integrated heating/cooling/hot water heat-pumps stack functions to save even more.

If you are willing to wash clothes in cold water and line-dry them (sunlight is a great disinfectant), you can increase your savings by $0.25 a load or more. For a family of four, that's $100 a year. Line-drying also lengthens the useful life of clothes. Even basement-dried clothes can be fluffed up in a dryer and still save you money. To further soften clothes, add a little baking soda or vinegar (not both!) to the wash.

Now that you have sharply reduced your utility bills, you can save much faster for a high-efficiency washer, refrigerator, windows, heat pump, furnace and the like. Some Energy Star-certified washing machines can cut water use by a factor of six and energy consumption by a factor of three. Because these upgrades substantially reduce monthly expenses, lenders are beginning to offer Energy-efficient mortgages to cover their cost at the time of a home purchase.

If you rent, your landlord may finance all or part of these upgrades. It doesn't hurt to ask. In many states in the USA, tax credits and other incentives can make energy upgrades affordable; you can look them up at www.dsireusa. org.

If you own a house, you needn't stop at appliance upgrades. In a cooler climate, a greenhouse extension along the sunny side of your house can reduce heating bills and extend your food growing season. Just be sure to vent it well and plan for full or partial shade in summer. In any climate, you can install a metal roof to catch rainwater for use in your garden. To be truly effective against drought, you will need a very large cistern. Although it can be made inexpensively out of ferrocement, it may not ever pay for itself in reduced water bills. However, it will reduce your contribution to run-off, erosion and fossil aquifer depletion. As more communities are affected by contaminated water supplies, catchment will come to be a more popular source of drinking water. Be aware, however, that in some jurisdictions, you may not legally own the rainwater that falls on your property.

If you are very rich, it can be hard to reduce your footprint without losing significant status. Chalk it up to conspicuous consumption, a term that is rightly considered insulting. However, your staff may appreciate some assistance with these upgrades in their own homes. Any effort you make to take in boarders and otherwise reduce your own footprint will elevate your status with Gaia.

Travel Less and Lighter

The average American car travels about 20,000 km every year, consuming as much energy as a trip to orbit aboard the single-stage rocket described in Chapter 2. Unlike our hydrogen-powered rocket, however, a car also generates over two tonnes of carbon dioxide pollution. High-efficiency hybrids produce perhaps half as much pollution per kilometer, but building them still produces about four tonnes of CO_2, just like any other car.

The upward spiral in car ownership forces towns, cities, developers and nations to pave over large areas of land to make room for ever more roads, parking and bedroom communities. As these efforts fill in wetlands and clear forests, they cut the remaining regenerative habitats into ever smaller areas. This not only diminishes Gaia's resilience to natural and artificial stresses, but also human resistance through road rage, isolation and alienation.

The quickest way to break this cycle is to live where you don't have to own a car. You may not find such an opportunity overnight, but it's worth watching for. Walkable towns and cities make for more and better friendships, and leave more money in your pocket. Without auto loan payments, fuel, insurance, parking, taxes and maintenance, you can easily save $10,000 a year.

In the meantime, you can reduce your transportation footprint by walking or bicycling as much as possible. When you need motor transport, a good rule is to plan it so that you never ride alone. Use public transportation where available, coordinate errands with friends and family, car pool and go online to find ride shares for long-distance drives. Not only does carrying three passengers reduce your own transportation footprint nearly fourfold, it also expands your social network.

Our car-choked towns and the spaces between them don't need more roads, they need more buses. If a tenth of existing car lanes were dedicated exclusively to bus transport and fully utilized, a significant fraction of the remaining roadways could be returned to the commons as parks, greenways, pedestrian sidewalks and bicycle paths. All while accommodating substantial population shifts and surges.

The situation faintly resembles South Africa in the days of apartheid. The haves (car-owners) bitterly resist any notion of sharing with the have-nots (the growing ranks of the poor and countless species now facing extinction due to decimated habitats). Ironically, all of us "haves" ultimately depend on the services of nature for our survival. Our transportation infrastructure comes

Vehicle	Footprint	Vehicle	Footprint
Car	100	Bus, in-town	40
Airplane, 1st class	60	Train, inter-city	40
Airplane, economy	50	Train, metro	25
Car, hybrid	50	Bus, inter-city	10

Table 8.1: Carbon footprints in kg CO_2 produced per 1,000 km per vehicle (cars) or per passenger (all other modes of transport, assuming typical usage patterns).[4]

directly at the expense of nature's life-support system. So it is we who are being subsidized non-renewably.

Heavy consumers tend to sneer at buses as a last-resort solution for the poor. Yet high-quality, inexpensive lines exist: I have especially fond memories of riding Honolulu's surf rack-equipped public *BeachBus* in the 1970s, and San Francisco's yacht-like *Green Tortoise* down from Seattle in the 1990s.

While you're waiting or perhaps agitating for your town to develop, improve and link its pedestrian, bicycle and bus infrastructure, consider more immediate ways to avoid owning a car. Perhaps the most successful strategy to date is car-sharing cooperatives, which have grown to over a hundred thousand members in the USA in recent years.[242] Car sharing can reduce car ownership sixfold or more while filling in the gaps in other modes of transit. It can also cut your automotive expenses including loan payments, insurance, gas and parking by a factor of up to twenty, depending on how often you drive.

What about other modes of transport? Virtually any alternative produces less carbon dioxide exhaust (a good measure of overall environmental impact) than a car with a single passenger. You can use Table 8.1 to weigh the carbon output of various vehicles. For example, if you drive a hybrid car, which produces 50 kg of CO_2 per 1,000 kilometers, you could carry a passenger and cut your emissions down to around 25 kg/1,000 km—about the same as you would produce riding metro train. To match a long-haul bus, you would have to cram 5 passengers into your hybrid without losing any fuel economy. Bicycles do even better: for typical daily commutes (few km) and diets (too many calories), cycling has no measurable effect on food demand, hence a negligible carbon footprint. Long-haul bicycling produces 2 kg of carbon per 1,000 km.[243] In other words, a bicycle costs you and the environment 50 times less than a car.

Recycle and Renew

We cannot achieve a zero-waste society until we close all of our resource loops. Ultimately, we must recycle *everything*.

My town of Bloomington, Indiana has a fairly progressive trash policy: once-a-week pick-ups and you pay $2 per bag. Most of the town also has curb-side recycling pickup every other week with two bins: one for mixed paper (including cardboard, newspaper, catalogs, magazines and office stationary), the other for metals, seven grades of plastic, and glass. If your town has a system like this, use it; if not, phone city hall and ask for it.

Kimberly and I put out a standard-sized can of garbage once every four to six weeks. A neighboring 3-person house puts out two overflowing cans every week. Why the difference? Protruding out of our neighbor's bins is a profusion of pizza boxes, plastic containers and other fast-food packaging. In contrast, most of our food is fresh produce, and we compost most of our kitchen waste. We re-use our own cloth grocery bags and recycle nearly all of our paper, glass, plastic and metal. There's not much left to throw away.

Our local recycling center has an area where people can leave books and household goods for free pickup and re-use. If your town doesn't have something like this, you can still go online and use freecycle.org to find or give away or find free items; or browse craigslist.org for free, barter and cash offerings.

I try to do as much of my work as possible on the computer (a laptop, of course: it uses fewer materials and consumes less energy than a gigantic desktop machine). Still, I end up buying a ream of paper a year. At major office supply stores, 100% recycled can cost nearly twice as much as the cheapest grades, but to me it's well worth it: a significant fraction of non-recycled paper, including papers used in catalogs and facial tissue, comes from liquidating the few remaining ancient forests. [244]

The introduction to Greenpeace's new deforestation map reports that the past 10,000 years have seen the gradual loss of 90% of Earth's intact forests to logging, burning, settlement, and fragmentation due to road-building. [245] Half of this loss occurred during the last 80 years, and half of that since the mid-1970's. In other words, our exponential growth is coming at the direct expense of the forests. Is there a viable alternative?

Yes. When you buy anything made of paper or wood, there are two easy ways to tell whether it came from the regenerative economy. For paper products, look for the words "recycled, 100% post-consumer content." Merely replacing your family's toilet paper with a brand that uses recycled rather than virgin pulp can save an old-growth tree every five years.[244] Second, for lumber and some office papers, look for the FSC logo. That stands for "Forest Steward-ship Council," an organization that audits lumber and paper product companies for sustainable forestry practices. The certification means that the wood used in the product was not clear cut old-growth, but instead was grown and harvested using methods that maintain the integrity of the forest ecosystem.

Even so, conservation is the first ethic of ecology. I minimize my forest im-pacts by using a handkerchief rather than facial tissue, cloth towels rather than paper towels and fabric napkins rather than paper napkins. I have also lived most of my life in very small dwellings, substantially reducing my demand for lumber.

Become Financially Independent

Did this section title surprise you? Thus far, we have been talking about con-suming less, the very opposite of the opulent lifestyle many of us associate with financial autonomy. Yet according to Joe Dominguez and Vicki Robin-son, authors of *Your Money or Your Life*,[246] frugality is one of the three pillars of wealth. The other two are earnings (of course) and getting out of debt.

Long-term financial debt such as credit cards, car loans and even mortgages promotes a lifestyle of waste. By encouraging you to spend what you have not earned, it promotes consumption, externalizes ownership and weakens local ties. It's bad for you and the environment. Dominguez and Robinson's solu-tion: cut your expenses everywhere possible and pay down the debt as fast as you can. The authors believe that on their plan, you can become financially independent in seven years. After that, you can say goodbye to long commutes and expensive business wardrobes. You can fill each day with family, friends and hobbies. You can become more rooted, more connected, more involved in your community. All of which shrinks your ecological footprint.

Embrace the Solar Economy

Going native to your local ecology involves becoming less of a consumer, as we've seen, but also more of a producer. Unlike extraction, production in the regenerative economy confers multiple yields: to yourself and your family, to the soil and to Gaia.

Our industrialized society is so predicated on competition that it is hard to even recognize such a win-win game. We tend to see our own egocentric consumer culture everywhere we look. At the cosmic scale, it becomes the Space Frontier. On the planetary scale, it becomes a zero-sum competition between individual species. However, biologists such as Lewis Thomas have long-since noted the amazing level of cooperation among species down to the sub-cellular level.[7] Most new species emerge not by taking a niche away from others, but by populating new niches that arise in an ever-changing environment and ecology. In other cases, these niches are not much more than opportunities for beneficial symbiosis (like the opportunity that gave rise to our mitochondria). To most people, evolution means survival of the fittest, but the evidence just as well supports another conclusion: that species evolve primarily to better serve all of life.

You can enjoy life's regenerative and cooperative genius more by becoming a producer in the solar economy.

Tend Your Own Garden

One of the easiest ways to begin is to grow some or all of your own food. There is enough room in the lawns, yards and patios of America's residences to feed much of the nation on Dr. Fuhrman's diet. As we'll see, you don't even need to own a yard to become a grower.

It's hard to come away from a friend's natural garden without a handful of produce, cuttings or seeds. Organic gardeners tend to be generous. Perhaps it's because they directly experience the bounty of living ecosystems. The more gardeners work with nature rather than against it, the more giving the relationship becomes for everyone. Even the most harried greenhouse grower I've met describes her sleepless springs not as battles but as caring for thousands of needy babies. Natural gardening is the opposite of war.

If you have gardened using chemical fertilizers and pesticides, switching to organic methods won't be that hard. If you haven't gardened, I would recommend starting immediately by sprouting seeds and grains at home for use in salads and sandwiches. Then you can move on to potted herbs and, at your earliest convenience, join an organic gardening club or take a course on the topic. When it comes to working in natural soils, I'm struck by how difficult even the basic things are to a newbie like myself, and how easy they seem to anyone with direct experience. In a garden, a lot of things are beyond your control. Plants die. Weeds and pests encroach. But many plants, such as lettuce and squash, volunteer new growth year to year without replanting. Often, plants will grow fine in less-than-ideal lighting and soils. Gardening develops attentiveness, experimentation and an accurate sense of humility.

One caution, though. If you wish to garden regeneratively, you need to make sure to use only open-pollinated, organic non-GMO (genetically modified organism) varieties. As much as possible, you should also plant species native to your region. In any case, your best source for seeds, cuttings and gardening tips is nearly always going to be local organic growers.

The primary barrier to self-sufficiency from a garden is the large area required for starch crops such as grains, beans and (to a lesser extent) potatoes and squash. If, however, you get most of your nutrition from greens and fruits per Dr. Fuhrman's guidelines, less than half of your growing area need be devoted to starchy plants.

As a producer, not only can you feed yourself and your family, you also may choose to produce and sell a surplus. From Pasadena to Philadelphia, Americans are growing food for profit in urban environments using organic intensive techniques. One such technique, Mel Bartholemew's square foot gardening, uses raised beds, close spacing and daily attention to achieve high yields throughout a prolonged growing season.[194,247] Permaculture techniques rely on beneficial guilds of edible plants, especially perennials, to achieve high yields with low maintenance costs. You don't even need to own any land to start growing: many neighborhoods are converting portions of public parks to community gardens for personal use. In many cities including Honolulu, people are starting to rent and profitably farm small plots on vacant lots, rooftops and neighboring yards.

Experience the Unthinkable

Most residential lots have poor soils, so organic gardeners are always looking for ways to improve it. One source that we have already mentioned is composting, that wholesome microbial activity that converts a compact pile of leaves, cut grass and kitchen waste gradually into excellent soil. Composting can range from throwing anything organic onto a garden bed to a carefully-orchestrated ritual involving giant thermometers, careful records, scheduled mixing, and Carbon to Nitrogen ratio calculations. No course or book on organic gardening would be complete without a discussion of compost, but the dirty little secret is that anything biodegradable rots away to compost over time.

That said, if we really want to close the resource loops, we need to go one step further and stop sending human excreta "away" in the form of sewage. Instead, it should go back into our gardens. Of course, we can't just use it in raw form: that's a recipe for cholera and other diseases. Where it belongs is in compost.

A few months ago, I directly experienced this unthinkable practice at a permaculture gathering. I arrived in need of a bathroom visit, and was shown the facilities: a low, boxy wooden cabinet with a normal-looking toilet seat on top. The lid was up and directions were clearly painted on its underside, but I was born with a bewildered expression. So my host helpfully pointed out where the sawdust bin was (it looked more like fluffy wood shavings than the powdery stuff I was expecting), and said "the goal is to use just enough so there is no standing water." Beneath the seat was a 20-liter bucket, about 1/4 full of sawdust. To my relief, there was absolutely no odor. During the work party that followed, I cleared some fencing from around a humanure compost pile that was three years old. It, too, was free of any odor other than that of healthy soil. Needless to say, the garden was thriving.

Humanure makes perfect sense in water-stressed suburbs: just add it to the compost heap, keep the heap covered with mulch and above flood zones, and heap's thermophilic bacteria will gradually kill the pathogens. [198] However, this practice remains illegal in most towns, cities and states in the USA.

Other home-scale waste recovery schemes do somewhat better than sawdust buckets at complying with local zoning, building and health codes. Commercial composting toilets with vents and sometimes fans and heaters process everything internally and may only need emptying once every few years. Aquatic systems such as Living Machines (livingmachines.com) purify blackwater (water-flushed sewage) in a series of miniature wetlands. Simpler and much smaller gravel-covered plantings can purify the greywater left over from washing and bathing. [115]

Wetlands in general have an undeservedly bad reputation in the USA as breeding grounds for mosquitoes and disease. That's a pity, because healthy wetland ecology supports predators such as minnows and dragonflies that control the mosquito population. Drained wetlands can have ten times greater mosquito densities than healthy ones because mosquitoes need just the occasional stagnant pool to breed, while their predators generally need more stable conditions.[248]

Throughout Europe, wetland ecology has become popular among homeowners, resorts and townships as a way to purify pond water to drinkable quality. Many owners of once-chlorinated swimming pools are now retrofitting them with shallows, aerators, and suitable natural guilds. While the pools and surrounding decks do have to surrender some area to nature, they also end their dependency on toxic, expensive chemicals. Natural swimming ponds essentially clean themselves.

Nature is already well-equipped to handle basic sanitation at the local scale. By letting small ecosystems manage our waste locally rather than sending it "away," we can improve our garden yields and overall quality of life.

Energy Farming

Pasadena's Dervaes family doesn't just produce food. Starting in 2004, they also began producing energy using an array of solar cells mounted to the roof of their home. They did the installation themselves and a rebate through the California Solar Initiative covered 2/3 of the cost of the panels. The array produces 7.2 kilowatt-hours (kWh) of electrical power on a good day. Using compact fluorescents, solar thermal water heating, Energy Star appliances and other conservation measures, the family reports a consumption level of 6.0 kWh per day. Through their utility company's net metering program, they are able to sell their excess power back to the grid and become a net energy exporter for part of the year—quite a distinction in a town plagued by summer power shortages.

The Dervaes also brew their own biodiesel fuel from waste vegetable oil. However, given that the average person may, at most, produce a few dozen liters of waste vegetable oil per year and that the average car burns 2,000 liters of fuel per year, post-consumer biodiesel is clearly not a solution for everybody.

In many areas, the best energy farming option by far is wind power. Large generators already compete economically with coal and nuclear plants; yet wind still accounts for only about a thousandth of the world's total power generation. Investors are moving fast on this: the total wind generation capacity grew tenfold between 1995 and 2004. Wind power can work at the neighborhood and even household scale. However, wind turbine power depends on the turbine's height above the ground, the square of the blade length and the cube of the wind speed, so size and location matter. In some locations, solar and wind systems can break even financially in less than ten years without subsidies.

Renewable Energy Credits

What if, despite all attempts at conservation and production, you still end up consuming more power than your home renewably generates throughout the year? It might not be a significant problem for the environment if your power company used wind or solar power. But according to the EPA's Power Profiler, non-hydro renewables such as solar and wind account for only 2% of U.S. power generation. In Bloomington, 91% of our power comes from coal, a polluting non-renewable. [233]

It would be good public policy to tax coal utilities and invest the proceeds in new wind energy projects—a clear example of the third Daly Rule (page 240). However, you don't have to wait for the government to enact this tax (which would of course be passed on to you, plus accounting overhead). You can, in effect, tax yourself via Renewable Energy Credits (RECs).

It works like this: first you calculate your non-renewable consumption, then you buy RECs sufficient to fund a renewable energy source that will replace that consumption over its lifetime. Because many combustibles such as coal, gas, oil and kerosene produce CO_2, a notorious greenhouse gas, it is convenient to express your impact in terms of your carbon footprint. The average American produces about 20 tonnes of CO_2 per year (five times the world average). [41] Not all RECs claim to offset your carbon footprint, but the ones that do cost around $12 per tonne of CO_2. The average American therefore should buy $240 worth of RECs a year to offset his or her consumption.

You can calculate your own carbon footprint at climatecrisis.net. Note that the calculator neglects the carbon impacts of home construction, manufactured goods, agricultural products and road infrastructure. Thus it makes sense to round up when you click the link to REC broker nativeEnergy.com to set up a payment plan.

Cottage Industry

Many people in America find it hard to extricate themselves from the consumer economy if for no other reason than holidays and birthdays. If you can get friends and family not to exchange gifts, count yourself lucky; most would regard such a request as pure blasphemy. This is one area where being a producer really helps: what can top a hand-made sweater, knit socks, or preserves made from back-yard fruit trees? DIY (Do It Yourself) is increasingly hip.

Let's assume the DIY trend continues and skip forward in time for a moment. As whole towns go native to their environment, demand for local-scale manufacturing will rise. For most urbanites, this conjures images of cottage industry complete with blacksmiths—a giant technological leap backward. But today's technology excels at embedding knowledge directly into machinery. It should be possible to automate high-precision manufacturing and recycling on the small scale. This would allow even small villages to produce their own custom clothing, furniture, electronics and building materials from feedstock that regenerates rapidly (such as bamboo), as well as fully recycled stocks of glass, plastics, aluminum and steel.

Small-scale production may seem to contradict the economies of scale that I argued would improve quality and lower prices in the case of the beverage-can rocket. However, the products we are considering here are not end-products such as shoes and mobile phones, but the machinery that would make them. If each village of a few hundred homes had a few bench-top fab and recycling outfits, demand for these devices would run well into the millions of units. As for the finished goods themselves, they would overcome the two primary *dis*-economies of scale inherit in large-scale manufacturing: inventory warehousing (none needed when everything is custom-made) and distribution (walking distance). Desktop fabs are currently under development at MIT.[156] To bring fabs into the regenerative economy, these devices need counterparts that recycle finished goods back into feedstock. That probably means redesigning most products so they can be recycled easily with zero waste and little energy expenditure, as suggested by William McDonough and Michael Braungart.[154] To be fully usable at the local scale, fabs also should be maintainable and repairable on-site.

Of course craft knowledge probably would fill the gaps in even the most sophisticated village fabs. Woodcraft, bicycle repair, spinning, knitting, weaving, electronics and other trades may not be in especially high demand now, but in the regenerative economy, they may very well be—and they're fun to learn.

This economic and cultural shift is already well under way: how-to media such as *Make* magazine and *Home and Garden* Television are thriving. Knitting has made a comeback; even American men are trying it. All of which gives hope for a smooth economic transition to full local production.

But *why* is DIY catching on in this era of global trade? Is it simply cooler to be a nerd, now that some of them have become famously rich? Perhaps, but when you ask a knitter in this land of the $1 felt cap why they would put $80 worth of their time into making one of their own, they'll typically answer that they like to see the results of their work. Come to think of it, I'm pretty sure that's why I baked pies around exam time back in graduate school. Unlike my course work, a pie was something tangible that I could make, enjoy and share in a couple of hours. When the globalized economy makes obtaining food, clothing and shelter too much like the driest kind of homework—bureaucratized and disconnected from direct experience—then making something yourself has an appeal that defies the strict logic of ledgers and receipts.

Accounting's precise numbers can confer a veneer of perfection that all too easily obscures the feel of real things. Therein lies a tragedy. When India began intensive trade with Ladakh, for example, the price of grain, once stable, became wildly unpredictable. In addition, vendors were pushing locals to buy pesticides, jersey cows and other unfamiliar products without full disclosure of their shortcomings and proper use. This had dire consequences for the Ladakhi, whose traditional form of exchange was mediated barter. In the days when they made or knew who made every finished good, they understood what went into it. Prices would fluctuate little, and when they did, perhaps due to drought, people would know the cause. [149]

You do not need to herd *dzo* to reap the benefits of cottage industry and trade. Quite the opposite. As you work more with what and who is native to your environment, the resources in your life become part of an observable cycle. Your actions and their effects become much clearer. Where life had been consumed with things to be extracted or thrown away, it now becomes regenerative, abundant, personal and comprehensible. Welcome to the solar economy.

Restore the Commons

As your daily decisions bring you ever more into the regenerative economy, you will find that further progress depends on how much you share with others.

The greatness of a civilization may be measured by the quality, accessibility and vitality of its commons: those areas of public life such as libraries, transportation, forests and lore that are cherished and enjoyed by all. We can become a better civilization not just by nurturing and protecting this domain, but also by remembering that the commons extend well beyond human use.

The greatest commons are the world's wild spaces, especially the forests, which do not charge for such services as carbon scrubbing, climate moderation, water purification and soil maintenance. However, human agriculture, road-building, waste disposal and other activities have removed 90% of the original forests, leaving soils in many areas damaged and vulnerable to erosion. Most of this destruction lurks hidden behind the smiling logos of modern commerce, but it sometimes pokes through in horrifying ways. When healthy forests are cleared, the land bleeds red, orange and brown. Where there's a buck to be made liquidating the commons, you'll see a lot of road kill.

Shrinking your footprint reduces demand for products (such as food and clothing) that come at the expense of the wild commons. This in itself helps prevent further destruction and allows damaged lands to heal. But recovery through strictly natural mechanisms can take centuries. Extinctions, of course, are permanent; when a region experiences too many of them, it can take tens of millions of years to regain its former biodiversity.

It is possible, however, to accelerate the pace of recovery by actively restoring damaged areas and protecting threatened ones from further abuse. Therefore, as frugality moves you toward financial independence, consider donating your surplus time and money to Gaia. You can, for example, volunteer for re-forestation and conservation work, or donate to the Nature Conservancy, which has protected 54 million hectares of threatened natural habitats world-wide, or to local conservancies that set aside wild lands in perpetuity.

However you become involved, though, I would caution you to seek ways to contribute that renew yourself as well as the land. Non-profits always need more money and labor, so you may find it helpful to use the permaculture principle of stacking functions to avoid burn-out. Look for restoration work that gets you outdoors in interesting places with knowledgeable people. Maximize your donations through such vehicles as gift annuities, which can earn you income.

Like the natural commons, the built commons also need active protection and development. If you try to ride a bicycle in the USA, you'll discover very few safe, legal routes. In most towns, every penny of transportation spending seems to have gone into roadways, not pedestrian and bicycle paths. This is ironic because a car requires nine times more road space per passenger per kilometer than a bicycle. If you try to take an intercity bus, which requires less than a twentieth as much roadway per passenger, you'll also find most areas poorly served. This is where pressure on neighborhood boards, city planning departments and the like can pay off over time. If you and your neighbors are already bicycling and car sharing, you'll have a better case.

Don't feel guilty if you don't know your neighbors very well: the gridded, car-focused streets that form the common spaces of most American neighborhoods strongly discourage social connection. To repair these broken commons, neighborhoods in Portland, Oregon have come up with a surprisingly effective tactic: painting local intersections. This brings neighbors together, draws out local artistic talent, declares their local identity and significantly calms traffic. For more information, visit www.cityrepair.org.

Cooperatives offer another good opportunity to develop the local commons. Beyond the community supported agriculture and car sharing co-ops that we've already discussed, your community may also have credit unions, utility co-ops, social and health services agencies, and consumer co-ops including health food stores and seed exchanges.

Perhaps the greatest of the built commons is the Internet. Beyond its obvious benefit of providing better access to information, it merges design, commerce and governance in a fluid, ever-evolving set of protocols and applications. Like any commons, it is vulnerable to abuse, which is why many of us are abandoning e-mail for applications that have better rules of participation such as discussion groups and wikis. By automating communications and record-keeping, the Internet puts powers of organization once reserved for nations and corporations into the hands of smaller groups and individuals. This can and has gone far toward supporting both locality and the development of large public bodies of arts, sciences and know-how. Local fabrication and recycling technology, for example, may soon advance primarily through open communities of machine operators swapping software and tips online.

It would be very helpful if everyone attempting to live regeneratively could share their experiences freely as the Dervaes have done. To facilitate this process, I have established a wiki at gaiome.org where you are invited to share your experiences and learn from others.

Summary: Becoming Cosmic Species

We are not yet the kind of civilization that can live for long anywhere in the cosmos, not even here on Earth. By embracing conquest, growth and waste, we have decimated the ecosystems that support us and millions of other species. By relying on extraction and consumption rather than regeneration, we never developed the skills it would take to build space-arks, nor the habits it would take to make them succeed.

To contribute to our survival rather than our destruction, your ecological footprint and mine must drop below a renewable 1.8 hectares. If you're an American, that's six times less than your neighbors: a hard number to achieve. If you have two or more children, your footprint needs to be smaller still.[4] Anything more than this leads directly to the resource wars that are engulfing our planet.

As we have seen, you can shrink your footprint immediately by owning little, living small, conserving energy, buying renewable energy credits, limiting travel and making local organic vegetables your primary food source. These may seem like acts of deprivation, but they also lead to financial security, closer friendships and substantially better health. By taking the next step and becoming a producer of food, energy and art, you can make nature's regenerative power a tangible part of your daily life.

Does living smaller mean turning away from science and technology? Not at all. If anything, it means more of it in the hands of more people, not just the elites. To thrive in the solar economy, we need to use available light, heat, wind and other natural flows with ever more care and consideration. To break the pattern of extraction and disposal, we need to re-tool our industrial economy for zero waste throughout the product life-cycle. To transition to a regenerative ecology, we need to recycle our entire resource stream as locally as possible. To learn from one another on the global scale, we need computing, communications and transportation infrastructures that are durable and non-polluting, yet also flexible. These are huge opportunities for technical entrepreneurship.

A sustainable footprint on Earth is still 18 times larger than the $1,000 \text{ m}^2$ assumed for gaiomes in Chapter 5. Even the Dervaes, who shelter and feed themselves on an impressive 330 square meters per person, still depend to some extent on a profoundly inefficient transportation, energy and communication infrastructure. Technology still needs to mature to the point where their solar panels, computer, bicycles, roof shingles and hundreds of other essentials come not from several continents as they do today, but originate and recycle within a few kilometers. In the meantime, I am confident that any group that buys land and starts trying to live on 1,000 m^2 per person today will fail horribly, as most utopias have in the past.

More likely and usefully, existing towns and cities throughout the world will shrink their footprints to renewable levels as more families and neighborhoods embrace the solar economy as the Dervaes have done. Industry will respond (as it is beginning to already) to newfound demand for regenerative, non-toxic technologies. As people refuse to wage war on nature and each other, they will eat better, walk and bicycle more and enjoy cleaner air and water. Their health and wealth will improve, not at someone else's expense but to everyone's benefit. Many, recognizing the natural wealth they enjoy, will feel like sharing. The cultural commons will grow. Strife and waste will decline. Our knowledge and connections to life will deepen. The land will recover. Extinctions will slow a thousand-fold, returning to their natural rate.

As renewable wind and solar power become the primary sources of energy, hydrogen may become an important way to store and buffer its fluctuating supply. According to EarthWatch founder Lester R. Brown, the existing natural gas infrastructure would need only minor upgrades to replace methane with hydrogen.[177] If the hydrogen comes from water, it stores renewable electricity as renewable propellants. Liquefied, hydrogen has about the highest exhaust speed of any rocket fuel, and it doesn't pollute the atmosphere.

In the solar economy, space exploration will become truly public: a venue for travelers and scientists, professionals and amateurs. Zero waste, by then second nature in all aspects of life, will allow millions of people to reach orbit safely. As village-scale manufacturing improves, spacers will apply it aloft, knitting proto-gaiomes first out of terrestrial feedstock, then asteroidal and cometary ores. Eventually, Gaia's seeds will dot the skies, and children will linger in twilight to count them.

We were never alone in the universe, especially not here on Earth. If we have faltered on the brink of space, it is because we somehow forgot that no conquest lasts in the circle of life—let alone beyond it. Only together, in guilds already far more than human, will we find the skills to become cosmic species. The Book is open; the choice is yours. What will the next page bring?

Notes

1. Gerard K. O'Neill. *The High Frontier: Human Colonies In Space*. Apogee Books, Burlington, ON, Canada, 3rd edition, 2000. Reprinted 1976 edition, with additional chapters by Freeman Dyson, John S. Lewis and others. The cylindrical *Island 3* design appears on color plate 2.

2. Gerard K. O'Neill. The colonization of space. *Physics Today*, 27(9):32–44, September 1974.

3. Stewart Brand, editor. *Space Colonies: a CoEvolution Book*. The Whole Earth Catalog and Penguin Books, Sousalito, CA and New York, NY, 1st edition, 1977.

4. Jim Merkel. *Radical Simplicity: Small Footprints on a Finite Earth*. New Society Publishers, Gabriola Island, BC, Canada, 2003.

5. Konstantin Tsiolkovsky. *The Science Fiction of Konstantin Tsiolkovsky*. University Press of the Pacific, Honolulu, HI, 1979. Trans. monographs originally published in Moscow between 1893 and 1935.

6. S. Morgan Friedman. The inflation calculator. www.westegg.com/inflation/, 2006.

7. Lewis Thomas. *The Lives of a Cell: Notes of a Biology Watcher*. Viking Press, New York, NY, 1974.

8. Manned Spacecraft Center. Apollo 8 onboard voice transcription as recorded on the onboard recorder (data storage equipment). Technical report, NASA, Houston, TX, January 1969.

9. E. O. Wilson. *Biodiversity*. National Academy Press, Washington, D.C., 1988.

10. Unless otherwise noted, times are from en.wikipedia.org/wiki/Geologic_age and en.wikipedia.org/wiki/Timeline_of_evolution, retrieved spring, 2007.

11. John S. Lewis. *Physics and Chemistry of the Solar System*. Elsevier Academic Press, Burlington, MA, 2nd edition, 2004. Good overview of asteroid origins, distribution and composition.

12. E. Belbruno and J. R. I. Gott. Where did the Moon come from? *The Astronomical Journal*, 129(3):1724–1745, March 2005.

13. Alex N. Halliday. Terrestrial accretion rates and the origin of the Moon. *Earth and Planetary Science Letters*, 176(1):17–30, 2000.

14. Katharina Lodders and Jr. Bruce Fegley. *The Planetary Scientist's Companion*. Oxford University Press, New York, 1998.

15. Vladimir I. Vernadsky. *The Biosphere*. Copernicus Books (Springer-Verlag), translated from ca. 1926 Russian text edition, 1998. P. 70 asserts that the mass of all living organisms is 10^{17} to 10^{18} kg; editor Mark McMenamin points out that the actual number is very uncertain, citing a modern estimate of the dry (i.e. just carbon) total including organic detritus of 7.5×10^{14} kg. Here, I assume a hydrated, living biomass of 5×10^{14} kg, an average cell diameter of 10^{-6} m, and cell density 1,000 kg/m^3 to get 10^{30} cells.

16. 10^{30} cells with mean lifetime 3,000 seconds, over a period of a billion years is about 10^{43} experiments.

17. Richard Leakey and Roger Lewin. *The Sixth Extinction: Patterns of Life and the Future of Humankind*. Doubleday, New York, NY, 1995. P. 45 shows the first 5 extinctions and their recovery periods.

18. Jonathan I. Lunine. *Earth: Evolution of a Habitable World*. Cambridge University Press, Cambridge, UK, 1 edition, 1999.

19. U.S. Census Bureau. Historical estimates of world population. www.census.gov/ipc/www/worldhis.html retrieved April, 2005.

20. U.S. Census Bureau. Total midyear population for the world: 1950-2050. www.census.gov/ipc/www/worldpop.html retrieved January, 2006.

21. Ernst Stuhlinger and Frederick I. Ordway III. *Werner von Braun: Crusader for Space*. Kreiger Publishing company, Malabar, Florida, 1994.

22. Christopher Lampton. *Wernher von Braun*. Franklin Watts, New York, 1988.

23. Willy Ley. *Rockets, Missiles, and Space Travel*. Viking Press, revised edition, 1957.

24. Konstantin E. Tsiolkovsky. *Selected works of Konstantin E. Tsiolkovsky.* University Press of the Pacific, Honolulu, HI, 2nd edition, 2004. Trans. papers from 1883 to 1935.

25. T. A. Heppenheimer. *Countdown: A History of Space Flight.* Wiley, New York, NY, 1997.

26. Clarence G. Lasby. *Project Paperclip: German Scientists and the Cold War.* Atheneum, New York, 1971.

27. Paul Grigorieff. The mittlewerk/ mittelbau/ camp dora mittelbau gmbh - mittelbau kz. www.v2rocket.com/start/chapters/mittel.html, November 2006.

28. White Sands Missile Range. V-2 rocket. www.wsmr.army.mil/pao/FactSheets/V2/v-2.htm.

29. Wernher von Braun. Crossing the last frontier. *Collier's Weekly*, 129(12):24–29, March 22 1952. This issue kicked off a 2-year series on space exploration that later appeared as *Across the Space Frontier* (1952) and two follow-on books.

30. Patrick Buchanan. Resolution to open a congressional investigation into the Arthur Rudolph case. Congressional Record: thomas.loc.gov/cgi-bin/query/z?r101:E24MY0-B292:, 1989.

31. Frederick Jackson Turner. The significance of the frontier in american history. *American Historical Association*, 1893.

32. Greg Klerkx. *Lost in Space: The Fall of NASA and the Dream of a New Space Age.* Pantheon Books, New York, 2004.

33. Lt. Col. G. W. Rinehart HAF/CX. Toward space war. *High Frontier: The Journal for Space & Missile Professionals*, 1(3):47–52, 2005. This serious non-fiction paper discusses privileged orbits, a mainstay of science fiction stories since Isaac Asimov wrote as Paul French.

34. en.wikipedia.org/wiki/Human_embryogenesis, retrieved summer, 2007.

35. U.S. Census Bureau. Historical national population estimates: July 1, 1900 to July 1, 1999. www.census.gov/popest/archives/1990s/popclockest.txt retrieved January, 2007.

36. U.S. Copyright Office. Annual report of the register of copyrights 2000. www.copyright.gov/reports/annual/2000/index.html, retrieved January 14, 2007.

37. U.S. Census Bureau. Statistical abstract of the united states. www.census.gov/compendia/statab/past_years.html. 1902, p. 453.

38. *U.S. Census Bureau*[37] 2006, p. 521.

39. Tyler Volk. *Gaia's Body: Toward a Physiology of Earth.* MIT Press, 2003.

40. International Energy Agency. World energy outlook. www.worldenergyoutlook.org/, 2004. p. 59 reports a world energy consumption of 10,345 million tonnes of oil equivalent (Mtoe) in 2002.

41. International Energy Agency. World energy statistics. www.iea.org/Textbase/stats/, 2004. p. 58 reports that 1 Mtoe $= 41,868 \times 10^{12}$J. Dividing the 2002 world energy consumption[40] by 3.15×10^7 seconds in a year gives a net consumption rate of 13.7×10^{12} watts.

42. Al Gore. *Earth in the Balance: Ecology and the Human Spirit.* Houghton Mifflin Company, New York, 1992.

43. Howard T. Odum. Limits of Remote Ecosystems Containing Man. *The American Biology Teacher*, 25:429–443, 1963.

44. T. R. Malthus. *On Population.* Modern Library, New York, 1960. First published in 1798 under the title: An essay on the principle of population.

45. Daniel E. Sullivan. Metal stocks in use in the united states. pubs.usgs.gov/fs/2005/3090, 2005. The total steel in use in the U.S. reported here as of 2002 was 4.13 billion tonnes. This includes all built structures, but not wastes sitting in landfill (835 million t). With a typical density of 7.86 t/m^3, the total volume would be 525 million m^3.

46. T.D. Kelly and G.R. Matos. Iron and steel statistics. In *Historical Statistics for Mineral and Material Commodities in the United States: U.S. Geological Data Series 140, pubs.usgs.gov/ds/2005/140/.* U.S. Geological Survey, 2005. The cumulative world production of steel reported here (1900-2002) added up to 5.26 times the U.S. production in the same time period. This accounts for virtually all steel in use throughout the world, an amount that would occupy a volume of 2.76 billion m^3, equal to a sphere of radius 871 m.

47. Edward F. Tedesco and François Xavier Desert. The infrared space observatory deep asteroid search. *The Astonomical Journal*, 123:2081, April 2002. Equation 2 gives 329,000 asteroids larger than D = 1,800 m in the Main Belt. *Lewis*[11] estimates that 8% of these (some 26,000) are M-type (metal) asteroids.

48. Frances Moore Lappé and Anna Lappé. *Hope's Edge: the Next Diet for a Small Planet*. Tarcher/Penguin Books, New York, 2003.

49. Donella Meadows, Jorgen Randers, and Dennis Meadows. *Limits to Growth: The 30-Year Update*. Chelsea Green Publishing, White River Jct., VT, 2004.

50. Jared Diamond. *Collapse: How Societies Choose to Fail or Succeed*. Viking Adult, 2004.

51. Sylvia Hui. Hawking: Space colonies needed. *Chicago Sun-Times*, June 14 2006.

52. Carl Sagan. *Pale Blue Dot: A Vision of the Human Future in Space*. Random House, New York, 1 edition, 1994.

53. Robert Zubrin. *The Case for Mars: The Plan to Settle the Red Planet and Why We Must*. Simon and Schuster, New York, NY, 1996.

54. Ben R. Finney and Eric M Jones, editors. *Interstellar Migration and the Human Experience*. University of California Press, Berkeley, CA, 1985.

55. American Museum of Natural History. National survey reveals biodiversity crisis—scientific experts believe we are in the midst of fastest mass extinction in earth's history. Press Release, www.well.com/user/davidu/amnh.html, April 20 1998.

56. Virginia Morell. The sixth extinction. *National Geographic*, 195(2):43–59, February 1999.

57. Chris D. Thomas et al. Extinction risk from climate change. *Nature*, 427(6970):145–148, January 8 2004. Using three separate methods, the study's 19 authors estimate that 18–35% of species alive today will become extinct by 2050 due to climate change (global warming) alone. Other impacts, such as habitat destruction by human settlement, add another 20–50% to estimated extinction losses.

58. Augustine of Hippo. *Confessions, 8.7.17*. James J. O'Donnell, cat.sas.upenn.edu/jod/augustine/Pusey/book08, 2006.

59. Jared Diamond. The worst mistake in the history of the human race. *Discover Magazine*, pages 64–66, May 1987.

60. Konstantin Tsiolkovsky. *Dreams of Earth and Sky*. Athena Books, Barcelona, 2004. Originally published in Moscow in 1895.

61. J. D. Bernal. *The World, the Flesh, and the Devil: An Enquiry into the Future of the Three Enemies of the Rational Soul*. Kegan Paul, Trench, Trubner, London, 1929.

62. David Abram. *The Spell of the Sensuous: Perception and Language in a More-Than-Human World*. Pantheon Books, New York, NY, 1996.

63. Richard Heinberg. *A New Covenant With Nature*. Quest Books, Wheaton, IL, 1996.

64. The historical NASA budget is not well-publicized, though the summary at en.wikipedia.org/wiki/NASA_budget generally agreed with figures quoted in the *Statistical Abstract*[37] and www.nasa.gov/about/budget/.

65. The *U.S. Centennial of Flight*[87] reports that 5,800 passengers flew in 1926. The Aircraft Crashes Records Office in Geneva, Switzerland (www.baaa-acro.com/, retrieved June, 2007) lists 42 air fatalities for that year, a 1 in 138 chance of dying per passenger. The *Statistical Abstract*,[37] (1980, pp. 670–671) reports that 1976 was an unusually safe year for airlines (only 45 fatalities), but that among the 668 million passengers who flew from 1975 through 1977, there were 863 fatalities. Thus, if you flew during that period of time, you had a 1 in 774,000 chance of dying in a fatal crash: 5,500 times better than your odds in 1926.

66. The LEO figure includes 900 m/s of losses incurred fighting gravity and atmospheric drag during ascent, plus the 60 m/s required to circularize at 375 km altitude and later de-orbit. See computations at *gaiome.com* for details.

67. Tsiolkovsky's *Investigation of World Spaces by Reactive Vehicles* (1903)[24] first noted the relationship between ΔV, the propellant's exhaust speed V_e, the rocket's initial mass m_i and its final mass m_f after combustion:

$$\frac{\Delta V}{V_e} = \ln\left(\frac{m_i}{m_f}\right).$$

This is the *rocket equation*.

68. Wayne Lee. *To Rise From Earth: An Easy-to-Understand Guide to Spaceflight*. Facts on File, New York, 1995.

69. Mara D. Bellaby. Rough soyuz landing blamed on malfunction. *Associated Press*, May 26 2003.

70. James Oberg. Consultant report: Soyuz landing safety www.jamesoberg.com/soyuz.html, March 19 1997.

71. Boeing Commercial Airplanes. Statistical summary of commercial jet airplane accidents worldwide operations 1959-2003. www.boeing.com/news/techissues, retrieved May, 2004.

72. National Transportation Safety Board. Ntsb/arg-04/01 annual review of aircraft accident data: U.S. general aviation, calendar year 2000. www.ntsb.gov/publictn/A_Stat.htm, retrieved June, 2004.

73. Peter Diamandis and Patrick Collins. Creation of an accredited passenger regulatory category for space tourism services. In *First STA Conference on Space Tourism*. Space Transportation Association, June 1999.

74. Karl Sabbagh. *Twenty-First Century Jet: The Making and Marketing of the Boeing 777*. Scribner, New York, 1996.

75. Leonard David. Total tally of shuttle fleet costs exceed initial estimates. www.space.com, February 11 2005.

76. Traci Watson. NASA administrator says space shuttle was a mistake. *USA Today*, September 27 2005.

77. Blue Origin. Draft environmental assessment for the blue origin west texas commercial launch site. ast.faa.gov/pdf/20060622_Draft_EA_As_Published.pdf, June 22 2006.

78. United Space Alliance. www.unitedspacealliance.com/about/facts.asp, 2006. reports a staff size of 10,000, not including suppliers.

79. Paul Floren. It's a hard day's night: An inside look into the life of a 747. *International Herald Tribune*, June 17 1997.

80. Boeing Commercial Airplanes. Prices www.boeing.com/commercial/prices/, November 2006.

81. Warren Berger. Hey, you're worth it (even now). *Wired*, 9(6), June 2001.

82. Space Adventures. www.spaceadventures.com, March 2005.

83. Associated Press. NASA wants space station publicity, but not this bad. *Houston Chronicle*, April 24 2001.

84. Miraslav Verner. *The Pyramids: the Mystery, Culture and Science of Egypt's Great Monuments*. Grove Press, New York, NY, 1997.

85. World Travel & Tourism Council. *Industry as a Partner for Sustainable Development: Tourism*. Beacon Press, London, 2002.

86. Allan Turner. Hotel gets a million stars for its view, not amenities. *Houston Chronicle*, October 28 2001.

87. U.S. Centennial of Flight. Air transport commercial aviation an overview. www.centennialofflight.gov/essay/Commercial_Aviation/Tran-OV.htm, retrieved Spring, 2007.

88. X Prize Foundation. History of the x prize. www.xprizefoundation.com/about_us/history.asp, June 2005.

89. U.S. Centennial of Flight. The 747. www.centennialofflight.gov/essay/Aerospace/Boeing_747/Aero21.htm retrieved Spring, 2007.

90. Airbus S.A.S. Product viewer. www.airbus.com/en/aircraftfamilies/productcompare/, retrieved June, 2005.

91. Eclipse aviation corporation secures fifth round of private equity funding for $87 million. www.eclipseaviation.com, July 30 2003. Press release.

92. Mark Wade. Shuttle. www.astronautix.com/lvfam/shuttle.htm retrieved June, 2007. Reports a program cost of $6.744 billion (1971), which, with inflation,[6] works out to $32.71 billion (2005).

93. Wiley J. Larson and James R. Wertz. *Space Mission Analysis And Design (SMAD)*. Microcosm, Inc., 1992. p. 734 notes that if the first unit off a production line costs T, then the cost of producing N units is $L = TN^B$, where

$$B = 1 - \frac{\ln((100\%)/S)}{\ln 2}.$$

S is the learning curve slope. I use a value of 85%, typical for production runs of 100 units or more, and compute only the incremental cost of the final unit off the line: TN^{1-B}.

94. Charles Ferguson. *High Stakes, No prisoners: A Winner's Tale of Greed and Glory in the Internet Wars*. Times Business Books, New York, NY, 1999.

95. Muhammed El-Hasan. Rocket man: Web entrepreneur focuses on aerospace in el segundo. *Daily Breeze*, March 26 2003.

96. Forbes.com. Company details: Northrop grumman, July 2005.

97. Jennifer Reingold. Hondas in space. *Fast Company*, 2(91):74, February 2005. About Elon Musk's SpaceX.

98. SMAD,[93] p. 752.

99. Paul Kallender. Rocket problems postpone Japanese spy satellite launch www.space.com/spacenews/archive03/spyarch_102003.html. *Space.com*, October 20 2003.

100. Anatoly Zak. Falling on a city near you: Dangerous spacecraft reentries. www.space.com/news/spacehistory/dangerous_reentries_000602.html, June 2 2000.

101. J.-C. Liou. Orbital box score. *Orbital Debris Quarterly News: www.orbitaldebris.jsc.nasa.gov/newsletter/newsletter.html*, 10(3):9, July 2006.

102. European Space Agency. Space debris. www.esa.int/spacecraftops/ ESOC-Article-fullArticle_par-40_1092735450198.html, September 2006.

103. SMAD,[93] p. 211.

104. Rocket Racing League. Specifications. www.rocketracingleague.com/x-racers_specifications.html, September 2006.

105. G. Harry Stine. *Halfway to Anywhere: Achieving America's Destiny in Space.* M. Evans and Company, New York, 1st edition, 1996.

106. NASA TM X-881 Apollo Systems Description Volume II: Saturn Launch Vehicles. NASA Marshall Space Flight Center, ntrs.nasa.gov/archive/nasa/casi.ntrs.nasa.gov/19710065502_1971065502.pdf, 1964. Retrieved January, 2007. Pages 19-5,6 give a usable propellant fraction of 91% rather than the more optimistic 92.7% that Stine reports.

107. Ed Regis. *Great Mambo Chicken and the Transhuman Condition: Science Slightly Over the Edge.* Addison Wesley, 1990.

108. Boeing Commercial Airplanes. Orders and deliveries. active.boeing.com/ commercial/orders/index.cfm?content=modelselection.cfm\&pageid=m15525, November 2006.

109. SMAD,[93] p. 668.

110. Buckminster Fuller. Interview. *Minneapolis Tribune*, April 30 1978.

111. D. R. Hitchcock and J. E. Lovelock. Life Detection by Atmospheric Analysis. *Icarus*, 7:149–159, 1967.

112. James Lovelock. *The Ages of Gaia: A Biography of Our Living Earth.* Norton, 1988.

113. George R. Williams. *The Molecular Biology of Gaia*. Columbia University Press, New York, NY, 1st edition, 1996.

114. Jill Neimark. A conversation with Tyler Volk. *New York Times*, August 11 1998.

115. Toby Hemenway. *Gaia's Garden: A Guide to Home-Scale Permaculture*. Chelsea Green Publishing Company, White River Junction, VT, 2000.

116. Michael Parfit. Powering the future. *National Geographic*, 208(2):18, August 2005.

117. Mathis Wackernagel and William Rees. *Our Ecological Footprint: Reducing Human Impact on the Earth*. New Society Publishers, Gabriola Island, BC and Stony Creek, CT, 1996.

118. Jane Jacobs. *The Nature of Economies*. Random House, New York, 2000. Jacobs makes the case that local diversity rather than global specialization leads to economic prosperity—a recurrent theme in permaculture.

119. Nathanial Ward. *On the Growth of Plants in Closely Glazed Cases*. J. Van Voorst, London, 1852.

120. PlantExplorers.com. The wardian age. www.plantexplorers.com/explorers/wardian-age.htm, 2005.

121. Kevin Kelly. *Out of Control: The New Biology of Machines, Social Systems and the Economic World*. Perseus Books Group, 1995.

122. Frank B. Salisbury, Josef I. Gitelson, and Genry M. Lisovsky. Bios-3: Siberian Experiments in Bioregenerative Life Support: Attempts to Purify Air and Grow Food for Space Exploration in a Sealed Environment Began in 1972. *Bioscience*, 47(9), October 1997.

123. Mark Nelson and Tony L. Burgess. Using a closed ecological system to study earth's biosphere. *Bioscience*, 43(4), April 1993.

124. John Allen. *Biosphere 2: The Human Experiment*. Penguin Books, Middelsex, England, 1991.

125. Eugene P. Odum. *Ecology And Our Endangered Life-Support Systems*. Sinauer Associates, Sunderland, MA, 1989.

126. Grand Prairie Friends of Illinois. www.prairienet.org/gpf/, April 2005.

127. Steinhart Aquarium and California Academy of Sciences. www.calacademy.org/aquarium, April 2005.

128. R. M. Wheeler, J. C. Sager, W. M. Knott, G. D. Goins, C. R. Hinkle, and W. L. Berry. Crop production for advanced life support systems—observations from the kennedy space center breadboard project. Technical Report TM-2003-211184, NASA, Kennedy Space Center, Florida, February 2003.

129. Roger Highfield. Trouble in the bio bubble. *The Age*, January 23 2004.

130. Faye Flam. Biosphere experiment results may show future of space travel. *Philadelphia Inquirer*, March 1 2004.

131. The spelling reflects the genetic rather than hierarchical relationship between members of each of life's top-level taxa.

132. John Allen. The purpose of the 1984 biosphere conference. In *www.ecotechnics.edu/1984b2.html*, Biospere 2, AZ, 1984. Space Biospheres Ventures.

133. Abagail Alling, Mark Nelson, and Sally Silverstone. *Life Under Glass: The Inside Story of Biosphere 2*. Biosphere Press, 1993.

134. Philip Elmer-Dewitt and Edwin M. Reingold. Getting back to earth. *Time*, 142(14), October 4 1993.

135. Robert Scarborough. My life as a designer of soils. *Whole Earth*, (96), Spring 1999.

136. B. Siano. Captain future's terrarium of discipline. *Humanist*, 52(2), March/April 1992.

137. Daniel Hawes. The people & the idiots: Growing up in the cult of the biosphere 2. Unpublished; details at danielhawes.com, 2005.

138. A. Toufexis. The wizards of hokum. *Time*, 138(13), September 30 1991.

139. Sharon Begley and Jeanne Gordon. In the desert, big trouble under glass. *Newsweek*, 123(16), April 18 1994.

140. DanVergano. Brave new world of biosphere 2? *Science News*, 150(20), November 16 1996.

141. Charles Cannon. Building blocks for an ecosystem. *BioCycle*, 38(2), February 1997.

142. Jane Poynter. *The Human Experiment: Two Years and Twenty Minutes inside Bisophere 2*. Thunder's Mouth Press, New York, NY, 2006.

143. Gary Taubes. Biosphere 2 gets new lease on life from research plan. *Science*, 267(5195):169, January 13 1995.

144. Fred A. Bernstein. Sprawl outruns Arizona's biosphere. *New York Times*, May 28 2006.

145. John Allen. Biospheric theory and report on overall Biosphere 2 design and performance during mission one (1991-1993). In *Linnean Society of London, Fourth International Conference on "Closed Ecological Systems: Biospherics and Life Support"*, April 10, 1996.

146. R. Buckminster Fuller. *Operating Manual For Spaceship Earth*. Aeonian Press, Mattituck, NY, 1969.

147. Patrick Whitefield. *Permaculture in a Nutshell*. Permanent Publications, Hampshire, England, 4th edition, 2005.

148. Martin C. Heller and Gregory A. Keoleian. Life cycle-based sustainability indicators for assessment of the U.S. food system. Technical Report CSS00-04, Center for Sustainable Systems, University of Michigan, Ann Arbor, 2000.

149. Helena Horberg-Hodge. *Ancient Futures: Learning from Ladakh*. Sierra Club Books, San Francisco, CA, 1991.

150. Michael Pollan. *The Botany of Desire: a Plant's-Eye View of the World*. Random House, New York, 2001.

151. Bill Mollison. *Permaculture: A Designers' Manual*. Tagari Publications, Tasmania, Australia, 2nd edition, 2004.

152. The power of community: How cuba survived peak oil. DVD, 2006.

153. Megan Quinn. The power of community: How cuba survived peak oil. *Permaculture Activist*, 59:33–37, Spring 2006.

154. William McDonough and Michael Braungart. *Cradle to Cradle: Remaking the Way We Make Things*. North Point Press, New York, NY, 2002.

155. Bruce Sterling. The dream factory: Any product, any shape, any size - manufactured on your desktop! *Wired*, 12(12), December 2004.

156. The Economist. Desktop manufacturing: helping poor-world inventors. economist.com/science/displayStory.cfm?story_id=3786368, March 23 2005.

157. W. C. Feldman, S. Maurice, A. B. Binder, B. L. Barraclough, R. C. Elphic, and D. J. Lawrence. Fluxes of fast and epithermal neutrons from lunar prospector: Evidence for water ice at the lunar poles. *Science*, 4(281):1496–1500, September 1998.

158. Donald B. Campbell et al. No evidence for thick deposits of ice at the lunar south pole. *Nature*, 443:835–837, October 19 2006.

159. Paul Spudis. Ice on the Moon. *The Space Review*, November 6 2006. www.thespacereview.com/article/740/1 retrieved June, 2007.

160. Audrey Delsanti and David Jewitt. The solar system beyond the planets. www.ifa.hawaii.edu/faculty/jewitt/papers/2006/DJ06.pdf retrieved February, 2007.

161. Carl Sagan. Planetary engineering on Mars. *Icarus*, 20:513, 1973.

162. Isaac Asimov. *The Martian Way and Other Stories*. Fawcett Crest, 1955.

163. Freeman Dyson. *Disturbing the Universe*. Harper & Row, New York, 1979.

164. John S. Lewis. *Mining The Sky: Untold Riches from the Asteroids*. Perseus Books Group, 1997.

165. Gregory Benford and George Zebrowski, editors. *Skylife: Space Habitats in Story and Science*. Harcourt, Inc., New York, NY, 2000.

166. Dandridge M. Cole and Donald W. Cox. *Islands in Space: The Challenge of the Planetoids*. Chilton Books, 1964.

167. Minor Planet Center. Provisional designations. cfa-www.harvard.edu/iau/lists/Desigs.html, September 28 2006.

168. Alessandro Morbidelli. Origin and dynamical evolution of comets and their reservoirs. http://arxiv.org/abs/astro-ph/0512256v1, 2005. Lecture notes from the 35th Saas-Fee advanced course in Switzerland and at the Institute for Astronomy at the University of Hawaii.

169. A. Morbidelli, W.F. Bottke, C. Froeschel, and P. Michel. Origin and evolution of near-Earth objects. In W.F. Bottke, A. Cellino, P. Paolicchi, and R.P. Binzel, editors, *Asteroids III*, pages 409–422, Tucson, 2002. University of Arizona Press.

170. Leonard David. First strike or asteroid impact? the urgent need to know the difference www.space.com/scienceastronomy/astronomy/nss_asteroid_020606.html, June 6 2002.

171. John S. Lewis. Personal interview. in2space.com, April 26 1998.

172. Don Yeomans. NEO discovery statistics. neo.jpl.nasa.gov/programs, June 3 2006.

173. *Lodders and Fegley*,[14] pp. 180, 277, 311, 317.

174. D. Yeomans. Small bodies in the solar system. *Nature*, 404:829, April 2000. Presents evidence that C-type asteroids may be wetter than we think.

175. Minor Planet Center. Forthcoming close approaches to the earth. www.cfa.harvard.edu/iau/lists/CloseApp.html, June 8 2006.

176. Columbia Accident Investigation Board. Public hearing. caib.nasa.gov/events/public_hearings/20030423/transcript_am.html retreived March 2, 2007. Statements of Robert F. Thompson, Shuttle Program Manager from 1970 to 1981, explaining why NASA quoted ticket prices as low as $118/lb to orbit in early 1970s dollars.

177. Lester R. Brown. *Plan B: Rescuing a Planet Under Stress and a Civilization in Trouble*. W. W. Norton, & Company, New York, London, 1st edition, 2003.

178. Amory Lovins with Carl Fussman. The energizer. *Discover*, pages 52–56, February 2006.

179. Theodore W. Hall. AIAA 9-4524, Inhabiting Artificial Gravity. In *AIAA Space Technology Conference*, Albuquerque, New Mexico, September 28–30 1999.

180. To achieve a gravity of g gees at the equator, a gaiome rotating with a period of T seconds must have a radius $R = 9.8g(T/2\pi)^2$ meters.

181. Space Settlements—A Design Study. Technical Report SP-413, NASA, Washington, D.C., 1977. Proceedings of the 1975 Summer Study held at Stanford University.

182. NASA SP-413[181] (p. 99) gives this formula for the hull thickness of a spherical gaiome with radius R:

$$t_h = \frac{P_0/2 + P_g}{\sigma_w - \rho_h R} R$$

where P_0 is the atmospheric pressure, P_g is the equivalent pressure of pseudogravity, ρ_h is the density of the hull, R is its radius and σ_w is the working stress of the hull. The weight of the column of soil of thickness t_s and density ρ_s at the equator is $P_g = 9.8g\rho_s t_s$.

183. Air in a spherical gaiome of radius R has density at distance r from the central axis of $\rho = \rho_0 e^{-ax}$ where ρ_0 is the density at the equator, $a = 9.8 \cdot 1.29/(2 \cdot 101325 \cdot R)$ and $x = R^2 - r^2$.

184. *NASA TM-413*,[181] p. 99: the applicable thickness of a toroidal gaiome of major radius R and cross-sectional radius r is:

$$t_h = \frac{(P_0/2)(r/R) + (P_g/\pi)}{\sigma_w - \rho_h R} R$$

and its diameters are $2r$ and $2(R + r)$.

185. Frances Moore Lappé. *Diet for a Small Planet*. Ballentine Books, New York, 2nd edition, 1975.

186. American Dietetic Association and Dietitions of Canada. Position of the american dietetic association and dietitians of canada: Vegetarian diets. *Journal of the American Dietetic Association*, 103(6):748–765, June 2003.

187. Nyssa S.R. Woods. Self contained off-world residential environment. Technical report, NASA Ames Research Center, 1997. Grand Prize winning entry in the annual space settlement design contest for high school students.

188. United States Department of Energy. DOE/EIA-0383 Energy Outlook 2005.

189. Apogee Instruments. What are the best ppf levels for optimum plant growth? www.apogee-inst.com/faq_solar.htm#Q4 retrieved spring, 2007.

190. Eugene A. Avallone and Theodore Baumeister III, editors. *Mark's Standard Handbook for Mechanical Engineers*. McGraw-Hill, New York, 9 edition, 1987. p. 12-124.

191. Space Resources and Space Settlements. Technical Report SP-428, NASA, Washington, D.C., 1977. Final Report of the 1977 Summer Study at NASA Ames Research Center, Moffet Field, California.

192. If gaiomians can handle a range of pseudogravities from g to $g - \Delta g$, they can live anywhere along the inner wall of a spherical gaiome between the latitudes

defined by $\arccos(1 - \Delta g/g)$. For a gaiome of radius R, this works out to an area of

$$A_i = 4\pi R^2 \sqrt{1 - (1 - \Delta g/g)^2}$$

This corrects a conceptual error in *NASA SP-413*,[181] p. 61 and *NASA SP-428*,[191] p. 42, which compute the surface area of the cylinder that connects the circles of latitude defined above—an underestimate, because the terraces/stories required to live at higher latitudes asymptotically reclaim the enscribed area of the sphere.

193. University of Delaware. UD to lead $53 million solar cell initiative. *UDaily*, November 2 2005. www.udel.edu/PR/UDaily/2006/nov/solar110205.html retrieved March 1, 2007.

194. Mel Bartholomew. *Square Foot Gardening*. Rodale Press, Emmaus, PA, 1981.

195. The World Conservation Union. IUCN red list – summary statistics for globally threatened species. www.iucnredlist.org/info/tables/table1, 2006.

196. *NASA TM-413*,[181] p. 108 provides a handy formula for radiation dose in rem/year, which I've modified thus:

$$dose = 16e^{-(\rho_h t_h + \rho_s t_s)/1.06}$$

to account for hull and soil densities ρ_h and ρ_s (t/m^3) and thicknesses t_h and t_s (m). The background radiation at sea level on Earth is around 0.5 rem/year.

197. Michael Belfiore. CSS Skywalker: the Five-Billion-Star Hotel. *Popular Science*, 266(3), March 2005.

198. Joseph C. Jenkins. *The Humanure Handbook*. Jenkins Publishing, 1994.

199. A gaiome with radius R rotating with a period of T and pseudogravity g gees has a rotational speed of $V = 2\pi R/T = 9.8gT/(2\pi)$ along its equator.

200. T. A. Heppenheimer. *Colonies in Space*. Warner Books, New York, 1977.

201. From the force-power relation $F = P/V$, a PIT tug with a mass of M tonnes and exhaust speed V_e seconds operating at P watts feels an acceleration of $P/(V_e M)$. To reach 1 km/s in a year, then, $P = 0.0317 \cdot V_e M$. From the rocket equation,[67] the propellant mass is

$$M_p = M(e^{\Delta V/V_e} - 1),$$

and the PIT will exhaust it in $M_p V_e \Delta V/(2P)$ seconds.

202. Again using the force-power relation, a solar sail face-on to the sun with reflectivity α (typically 0.9 for aluminum) and area A will feel a force $F = 2(S/c)\alpha A$, where S is the solar constant and c is the speed of light. Suppose it is in a circular orbit at 1 AU, where $S = 1380W/m^2$. When it is tipped θ degrees about its orbital axis, it feels a "lift" component of force along its orbit of $F = 9.2 \times 10^{-6}\alpha A \cos^2\theta \sin\theta$. This force is maximized at $\theta = 35.265°$. Thus, to reach 1 km/second in a year, a gaiome with a mass of M tonnes needs a sail with an area of at least $A = (8965 \cdot M/\alpha)$ square meters.

203. Joel Grossman. Solar sailing: The next space craze? *Engineering & Science*, (4):18–29, 2000.

204. Webster Cash. New worlds imager. NIAC presentation circa 2005 casa.colorado.edu/~wcash/Planets/new_worlds.pdf, retrieved June, 2007.

205. Integrating the air density [183] throughout a spherical gaiome gives total air mass

$$M = 2\pi\rho_0 \int_R^0 e^{-ax}\sqrt{x}\,dx$$

with solution

$$M = -4\pi\rho_0 \left(\frac{\sqrt{\pi} \cdot erf(R\sqrt{a})}{4a^{3/2}} - \frac{Re^{-aR^2}}{2a} \right)$$

which for a 2 km diameter gaiome, amounts to 95% as much as the approximation $M = 4\pi\rho_0 R^3/3$. For a 20 km gaiome, the exact calculation used in the text comes to 69% of the approximation.

206. Craig Covault. Bigelow's gamble. *Spaceflight Now*, September 27 2004.

207. A practicing aerospace engineer presented the underbid/lock-in/overrun cycle as standard space industry practice in a special lecture to my Spaceflight Engineering class (MAE 432) at Princeton University on April 6, 1987.

208. Jim Giles. Internet encyclopaedias go head to head. *Nature*, 438:900–901, December 2005. In a blind, side-by-side comparison of 50 articles from *Wikipedia* and *Encyclopædia Britannica*, the study found four serious errors each in both sources, and 162 minor factual errors, omissions or misleading statements in *Wikipedia*, vs. 123 in *Britannica*. *Encyclopædia Britannica* disputed these findings and asked for a retraction, but *Nature* refused.

209. SMAD,[93] p. 209.

210. Sloan Digital Sky Survey. Data release 4. www.sdss.org/dr4/, June 2006.

211. Steven J. Ostro. Delta-v for spacecraft rendezvous with all known near-earth asteroids. echo.jpl.nasa.gov, June 10 2006.

212. Nigel Wells. SIMONE NEO mission study. www.esa.int/gsp/completed/neo/simone.html, 2003.

213. C. Lee Dailey and Ralph H. Lovberg. The PIT MkV pulsed inductive thruster. Technical Report CR-191155, NASA, 1993.

214. Robert H. Frisbee and Ioannis G. Mikellides. The nuclear-electric pulsed inductive thruster (NuPIT): Mission analysis for prometheus. In *41st AIAA/ASME/SAE/ASEE Joint Propulsion Conference and Exhibit*, Tucson, 2005.

215. Henry Ford. *My Life and Work*. Garden City Pub. Co., Garden City, NY, 1922. www.gutenberg.org/etext/7213, retrieved 2006.

216. Mark Sonter. Near earth objects as resources for space industrialization. *Solar System Development Journal*, 1(1):1–31, 2001.

217. Charles L. Gerlach. Profitably exploiting near-earth object resources. In *International Space Development Conference*, Washington, D.C., 2005. National Space Society. gerlachspace.com/resources/ retrieved April, 2006.

218. Robert Zubrin. *Entering Space: Creating a Spacefaring Civilization*. Jeremy P. Tarcher/Putnam, 2000.

219. *Vernadsky*,[15] p. 62.

220. N. S. Kardashev. Transmission of information by extraterrestrial civilizations. *Soviet Astronomy*, 8:217, 1964.

221. *Lodders and Fegley*,[14] pp. 241: Ceres, with 1.2×10^{21} kg, is 1/4 of the total mass of the Main Belt. I assume the mass of undiscovered Trojans is equal to the mass of the Main Belt.

222. For gaiomes of cross-sectional area A separated by mean distance d, the time between collisions with average impact speed v is $t = d^3/Av$.

223. George Zebrowski. *Macro-Life*. Avon, 1981.

224. Carl Sagan, editor. *Communication with Extraterrestrial Intelligence (CETI)*. MIT Press, Cambridge, MA, 1973. Dyson's essay was later reprinted in Sagan and Druyan's 1985 book *Comet*.

225. R. L. Forward. Starwisp: An ultra-light interstellar probe. *Journal of Spacecraft and Rockets*, 22:345, May 1985.

226. While producing offspring does not in itself imply sexuality or gender, I refer to Gaia as a "she" anyway, to personalize her parental relationship to us and all living things on Earth.

227. H. J. Melosh. Exchange of meteorites (and life?) between stellar systems. *Astrobiology*, 3:207, 2003.

228. R. M. E. Mastrapa et al. Survival of bacteria exposed to extreme acceleration: implications for panspermia. *Earth and Planetary Science Letters*, 189(1–2):1–8, June 30 2001.

229. Curt Mileikowsky et al. Natural transfer of viable microbes in space 1. from Mars to Earth and Earth to Mars. *Icarus*, 145:391, 2000.

230. W. M. Napier. A mechanism for interstellar panspermia. *Monthly Notices of the Royal Astronomical Society*, 348(1):46, February 2004.

231. Chellis Glendinning. *My Name is Chellis and I'm in Recovery from Western Civilization*. Shambhala Publications, Inc., Boston, MA, 1994.

232. Herman E. Daly. Economics in a full world. *Scientific American*, 293(3), September 2005.

233. If you live in the USA, you can get a detailed report on where your electrical power comes from by visiting the Environmental Protection Agency's energy profiler (www.epa.gov/cleanenergy/powerprofiler.htm).

234. A.K. Chapagain and A.Y. Hoekstra. Water footprints of nations. Value of Water Research Report Series No. 16 1, UNESCO-IHE Instutute for Water Education, Delft, The Netherlands, 2004. www.waterfootprint.org/ retrieved February, 2007.

235. Corinne T. Netzer. *The Complete Book of Food Counts*. Bantam, Doubleday, Dell, New York, NY, 2nd edition, 1991. You must eat 6 times as many grams of spinach (p. 22) as beef (p. 539) to get the same amount of protein; UNESCO's online water use calculator[234] shows that a kilogram of beef takes 35 times as much water to produce as a kilogram of veggies.

236. Joel Fuhrman, M.D. *Eat to Live: The Revolutionary Formula for Fast and Sustained Weight Loss*. Little, Brown and Company, Boston, New York, London, 2003.

237. Joel Fuhrman, M.D. *Disease-Proof your Child: Feeding Kids Right.* St. Martin's Press, New York, 2005.

238. Jannequin Bennett. *Very Vegetarian.* Rutledge Hill Press, Nashville, TN, 2001. Introduction by Carl Lewis also appears online at earthsave.org/lifestyle/carllewis.htm (June, 2007).

239. Rich Pirog and Tim Van Pelt. How far do your fruit and vegetables travel? www.leopold.iastate.edu retrieved february, 2007, Leopold Center for Sustainable Agriculture, Iowa State University, Ames, IA, 2002.

240. U.S. Food and Drug Administration. FDA statement on foodborne e. coli o157:h7 outbreak in spinach. Press release: www.fda.gov/bbs/topics/NEWS/2006/NEW01489.html, October 12 2006.

241. U.S. Centers for Disease Control. Update on multi-state outbreak of e. coli o157:h7 infections from fresh spinach, october 6, 2006. www.cdc.gov/ecoli/2006/september/updates/100606.htm.

242. CarSharing Network. Where can you find car sharing in North America. www.carsharing.net/where.html, retrieved November, 2006.

243. Worldwatch Institute. Matters of scale—bicycle frame. www.worldwatch.org/node/4057 retrieved March, 2007. Reports that a car consumes 1,860 calories per mile, while a bicyclist consumes 35; also that roads can carry 10 times as many people by bicycle as by car.

244. Ancient forest friendly tissue papers. kleercut.net/en/forestfriendly, retrieved June, 2006.

245. Greenpeace. Roadmap to recovery: The world's last intact forest landscapes. www.intactforests.org, retrieved 2006.

246. Joe Dominguez and Vicki Robin. *Your Money or Your Life: Transforming Your Relationship with Money and Achieving Financial Independence.* Penguin, London, England, 2nd edition, 1999.

247. Mel Bartholomew. *Cash from Square Foot Gardening.* Storey Books, North Adams, MA, 1985.

248. Brent Ladd and Jane Frankenberger. Purdue extension water quality program wq-41-w: Management of ponds, wetlands, and other water reservoirs to minimize mosquitoes. www.ces.purdue.edu/waterquality/resources/mosquitoes1.htm retrieved February, 2007.

Index